ST. CLAIR TUNNEL

Rails Beneath the River

ST. CLAIR TUNNEL

Rails Beneath the River

Clare Gilbert

Stoddart

A BOSTON MILLS PRESS BOOK

Canadian Cataloguing in Publication Data

Gilbert, Clare, 1947-
  St. Clair Tunnel

ISBN 1-55046-045-5

1. St. Clair Tunnel (Sarnia, Ont. and Port Huron, Mich.)
— History.   I. Title

TF238.S25G55 1991   385é.312   C91-094194-7

Published in 1991 by
Stoddart Publishing Co. Limited
34 Lesmill Road
Toronto, Canada
M3B 2T6

A BOSTON MILLS PRESS BOOK
The Boston Mills Press
132 Main Street
Erin, Ontario
N0B 1T0

Edited by Noel Hudson
Designed by John Denison
Cover designed by Gillian Stead
Typography by Lexigraf, Tottenham
Printed at Ampersand, Guelph

The publisher gratefully acknowledges the support of
The Canada Council, Ontario Arts Council and Ontario
Publishing Centre in the development of writing and
publishing in Canada.

Cover:
A St. Clair Tunnel Company 0-10-0 emerges from the Sarnia Portal eastbound amid a
cloud of steam and smoke, circa 1900. Note the well-kept floral garden at the base of the
north slope.   – Lambton County Library

To my father, the late Kenneth H. Gilbert, who bequeathed me his interest in things historical.

American Association
for State and Local History
Award of Merit

Winners of the
Heritage Canada
Communications Award

Contents

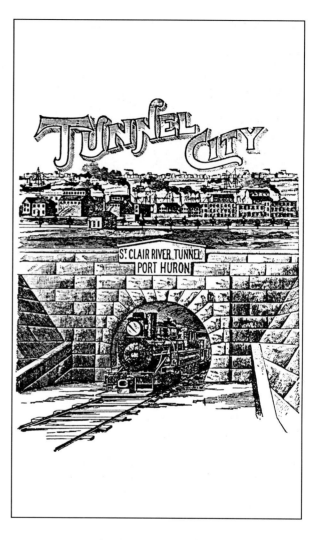

Cover of the Port Huron (Michigan) Times *special "Tunnel Opening" edition, September 19, 1891.*
– Port Huron Museum of Arts & History

The Lovat Earth Pressure Balance Tunnel Boring Machine was used to construct the new tunnel.

Foreword

In July of 1991, CN North America announced it would spend upwards of $1 million to examine the feasibility of making extensive improvements in the capacity of the 100-year-old St. Clair River tunnel that would permit it to operate double-stack container trains and multi-level auto carriers.

Following a detailed evaluation of proposals submitted to CN by a number of consulting engineers, Klohn Leonoff Ltd. was awarded the feasiblity study, which included a detailed environmental review.

Established in Vancouver in 1951, and with offices in major cities in Canada and the U.S., Klohn Leonoff and the project team have worked on more than 100 tunnel assignments throughout the world.

The feasibility study, completed in November of 1991 and approved by the CN Board of Directors December 3, 1991, recommended building a new tunnel, with an internal diameter of 8.3 metres and located approximately 30 metres upstream of the existing tunnel. The Board also approved adopting the "NCP," a Negotiated Compressed Process, in order to advance the completion date of the new tunnel by one year.

The NCP included prequalifying tunnel-boring machine manufacturers and tunnel construction contractors on a world-wide basis. Lovat Tunnel Equipment Inc. of Toronto, Ontario, was chosen to supply the first North American-made "Earth Pressure Balance" tunnel-boring machine.

The Lovat TBM excavated a diameter of 9.5 metres.

The new tunnel was completed in 1994.

Number 600 was renumbered 1303 in 1898. The side tanks are now gone; the cab has been sheathed with steel and the oil headlights replaced.

– Gary Shurgold/Bill Glassco Collection

Introduction

Stop anyone on the streets of Sarnia or Port Huron today and ask directions to the St. Clair Tunnel, and chances are you will be met with a blank stare. Today's citizens have little or no daily contact with the tunnel and the railways that use it. Such was not always the case. A hundred years ago Sarnia and Port Huron were in the grip of "tunnel fever." Every newspaper carried word of progress during construction, and the cities virtually closed down to celebrate the opening ceremonies. Entrepreneurs eagerly tacked the name "Tunnel City" onto a variety of goods produced in the two cities, these ranging from bicycles to stoves. One enterprising real-estate developer, C.F. & E.B. Harrington, even advertised an auction of lots in the "Tunnel Depot Subdivision" of Port Huron.

But things were different a hundred years ago. Railways were more than a way of moving goods, they were a vital means of communication and travel. They connected people with the outside world. The ordinary citizen was more intimately involved with the day-to-day happenings of the railway. A job on the railway was a status symbol. It was not uncommon to see announcements in the local newspaper that Miss So-and-So of Sarnia was to marry Mr. So-and-So of the Grand Trunk Railway.

The St. Clair Tunnel, which opened for traffic in 1891, was and still is an engineering wonder. Engineering trade journals from around the world marvelled at the bore driven through soft clay beneath the river bed. In the 100 years since its construction, the historical, engineering and financial importance of this tunnel has been overshadowed by more recent inventions, but there are few engineering works that have lasted 100 years and served their original purpose with such little change. As we approach the centennial of its completion, the St. Clair Tunnel remains as important an international link in the Canadian National Railways system as it was to its predecessor, the Grand Trunk.

One of a deck of playing cards produced by the Grand Trunk, circa 1910. Each card showed a different scene along the Railway.
– Author's Collection

9

The original Grand Trunk Station at Point Edward, constructed in the late 1850s. It burned down in 1871 and was replaced by a second, larger station. — NAC

CHAPTER ONE:

The Railways Arrive

In the late 1850s Sarnia, Ontario, was a beehive of railroad activity. The Great Western Railway of Canada built a branch of its Niagara Falls–Windsor main line from London in 1858 and constructed a waterfront depot at the foot of Cromwell Street. The following year, 1859, the Grand Trunk Railway of Canada completed its line from Portland, Maine, by way of Montreal and Toronto. For their terminal, the Grand Trunk chose a site north of Sarnia at what is now the village of Point Edward. Both of these railways were built to the provincial gauge of 5 feet 6 inches. In November 1859 the Grand Trunk extended their line from Fort Gratiot, Michigan, opposite Point Edward, to Detroit by leasing the standard-gauge Chicago, Detroit & Canada Grand Trunk Junction Railway. Connections were made with lines to Chicago at Detroit.

Docks were built at Point Edward and Fort Gratiot, and the Grand Trunk began a "break bulk" system of transferring freight, using the side-wheel steamer *W.J. Spicer* and other boats. This arrangement was time-consuming and extremely labour intensive. Freight had to be unloaded, transferred to a boat, and then reloaded on railway cars on the opposite shore. While the Grand Trunk's terminals were located at the narrowest point of the St. Clair River, it was also the swiftest point, with a current of about 8 knots. This created problems. Shipping was concentrated at this crossing and the Grand Trunk ferries had to dodge the lake boats. Also, during the winter months pack ice would drift down from Lake Huron, fouling the slips and

making navigation difficult. To overcome this tedious system, the Grand Trunk put into service a swing ferry capable of handling standard-gauge cars. This unique vessel was a barge anchored by a chain approximately 1,000 feet long and propelled by the current. By changing the anchor chain's angle of attachment, the current could be used to drive the ferry from one shore to the other.

In 1872 the swing ferry was replaced by the steam-powered car ferry *International*. Her iron hull was built by Palmer and Company at Yarrow-on-Tyne, England, then knocked down and shipped to Fort Erie, Ontario, where it was reassembled by the Grand Trunk. She was the first Great Lakes ferry equipped with three tracks and, at 210 feet in length, was capable of carrying 21 cars. Powered by two high-pressure, non-condensing engines, 30 inches by 30 inches, built by E.E. Gilbert & Sons of Montreal, she was driven by twin screws, making her the first propeller car ferry on the Great Lakes. Between 300 and 400 cars per day would be ferried on a normal day. The Grand Trunk began changing the gauge of its track from the 5-feet-6-inch provincial gauge to 4-feet-8½-inch standard gauge the same year, making interchange with its American line much easier.

By 1874 the Grand Trunk's Chicago–New England traffic had increased to the point that a second ferry was needed. Again the Grand Trunk ordered an iron hull constructed in England and had it shipped in sections to Point Edward, where it was assembled under the supervision of John

IR FERRY INTERNATIONAL FOR MANY YEARS FERRIED R.R. CARS FOR THE GRAND TRUNK RR
ETWEEN PT. EDWARD AND FT. GRATIOT, AFTER THE ST. CLAIR TUNNEL WAS COMPLETED
HE WAS SOLD TO THE PERE MARQUETTE R.R. AND SHE RAN BETWEEN SARNIA AND
T. HURON. LATER BETWEEN WINDSOR AND DETROIT AND FINALLY SOLD TO SOREL
INTERESTS.   POINT EDWARD 1890

The car ferry International *serviced the Grand Trunk's Point Edward–Fort Gratiot
crossing from 1872 until the tunnel was opened in 1891.* — Lambton County Library

The Huron *docked at Point Edward. Fort Gratiot Station, which still stands, can be seen in the background.*
— Port Huron *Times-Herald*

Smith. The *Huron* was launched June 3, 1875, and went into service between Fort Gratiot and Point Edward on July 1, 1875. With a length of 238 feet, she had a capacity of 24 cars, but the capacity of car ferries is deceiving. *International* and *Huron* were originally listed at 21 and 24 cars respectively. The fact that car sizes increased over the years meant that, while still capable of carrying the same tonnage, the boats moved fewer cars. This is reflected in the fact that in later years *Huron* was listed as carrying 11 cars. Even with this extra capacity the Grand Trunk was hard pressed to keep up with the transfer of freight across the St. Clair River.

Competition was developing for this traffic from the American Midwest to the Eastern Seaboard. Up to this time the Grand Trunk had made connections to Chicago by way of the Michigan Central. In 1876 William Vanderbilt gained control of the Michigan Central plus a 49-mile stretch of track between Flint and Lansing, the Chicago & Northeastern. Vanderbilt raised his rates such that the Grand Trunk was left with no viable alternative for reaching Chicago. Fortunately for the Grand Trunk, this also had the effect of drying up traffic on the short lines that connected with Vanderbilt's Chicago & Northeastern, and the Grand Trunk was able to acquire these lines. They convinced Vanderbilt to sell by threatening to build a line parallel to the Chicago & Northeastern. Vanderbilt had no choice, and by 1879 the Grand Trunk controlled their own lines into Chicago. On April 7, 1880, these lines were consolidated into the Chicago & Grand Trunk Railway. The Grand Trunk route between Chicago and Portland was now the world's longest railway under one management.

In 1882 the Grand Trunk absorbed its competitor, the Great Western. This merger included the Detroit, Grand Haven & Milwaukee Railroad, which ran between Detroit and Grand Haven, and connected with Milwaukee by steamer. This added traffic put a strain on an already overburdened crossing at Fort Gratiot. By 1883 the Grand Trunk was carrying one third of all traffic between Chicago and New England. After their victory over Vanderbilt, the Grand Trunk captured a large share of the grain trade by offering rates that the Vanderbilt-controlled lines had refused. The dressed-meat trade from the Chicago packing houses became an important source of income for the Grand Trunk, ranking third behind grain and livestock. This lucrative trade, with its requirements for icing and special handling, could not entertain the possibility of a delay. As the St. Clair River crossing was susceptible to the vagaries of winter weather conditions and the busy Great Lakes shipping traffic in summer, it became imperative that the Grand Trunk find a solution to these problems. A bridge or tunnel was the only answer. The heavy navigation at this narrow point in the river ruled out a drawbridge as being not much better than the ferry service. A high-level bridge, with enough clearance above the water to allow ships to pass underneath, would involve such long approaches as to be impractical. A tunnel was the only choice left.

As early as 1882 Sir Henry Tyler, president of the Grand Trunk, had anticipated the possibility of constructing a tunnel and had hired Montreal engineer Walter Shanley to do a study. Shanley's study recommended a site approximately three miles further downstream as being the most suitable loca-

Point Edward Yard in the 1880s. The constant backlog of freight, evident in this view, was the reason the Grand Trunk built the St. Clair Tunnel.

– Port Huron *Times-Herald*

The ferry dock was replaced by these ore unloaders. Iron ore from Minnesota was transshipped by the Grand Trunk to Hamilton before the opening of the current Welland Canal. The footings can still be seen just south of the Blue Water Bridge. — NAC

tion at which to drive a tunnel under the river.

It was no coincidence that this site was directly in line with the Chicago & Grand Trunk line on the U.S. side and the newly acquired Great Western line on the Canadian shore, which became GTR's main line across Southern Ontario. This resulted in saving approximately six miles of main line trackage through the centre of Sarnia and Port Huron had the tunnel been built at the site of the Point Edward–Fort Gratiot ferry crossing.

The St. Clair Frontier Tunnel Company was incorporated on April 19, 1884, in the Dominion of Canada, under Act 47, Vic. Cap 82, to construct, maintain and operate a tunnel between Sarnia, Ontario, and Port Huron, Michigan, with power to amalgamate with a similar company chartered in the United States. The Port Huron Railroad Tunnel Company was incorporated under the laws of the State of Michigan on October 18, 1886. The St. Clair Tunnel Company was created from the amalgamation of these two companies on November 9, 1886. This amalgamation was certified in Canada on November 15, 1886, and in Michigan on November 23, 1886. The officers of the company were Sir Joseph Hickson, president (later replaced by L.S. Seargeant); Robert Wright, secretary-treasurer; John Bell, E.W. Meddaugh, W.J. Spicer, Charles MacKenzie and Alexander Vidal, directors.

First-mortgage 5-percent bonds were issued in the amount of $2,500,000. These were acquired by the Grand Trunk with an issue of 4-percent consolidated perpetual debenture stock. This was the largest mortgage issued in the State of Michigan to that time.

The urgency with which the Grand Trunk regarded the tunnel becomes evident in light of the facts that the railway had never been particularly profitable and that a previous tunnel attempt between Detroit and Windsor, begun jointly by the Michigan Central and Great Western in 1872, had been a total failure.

To head the project, Tyler chose Joseph Hobson, chief engineer of the recently acquired Great Western Railway. Hobson was born at Guelph, Ontario, and served his apprenticeship as a provincial land surveyor. In 1870 he was appointed engineer of the International Bridge which was being constructed between Black Rock, New York (near Buffalo), and Bridgeburg, Ontario (near Fort Erie). Upon completion of that project in 1873, he moved to the Great Western Railway as assistant chief engineer and was later promoted to chief engineer. When the Grand Trunk gained control of the Great Western in 1882, Hobson assumed a similar position on the Great Western Division. Tyler picked the right man for the job. Hobson's combination of daring, tenacity and engineering knowledge was what a project of this magnitude needed.

Joseph Hobson, chief engineer on the St. Clair Tunnel project, was born at Guelph, Ontario, on March 4, 1834, and died at Hamilton on December 19, 1917. – CNR

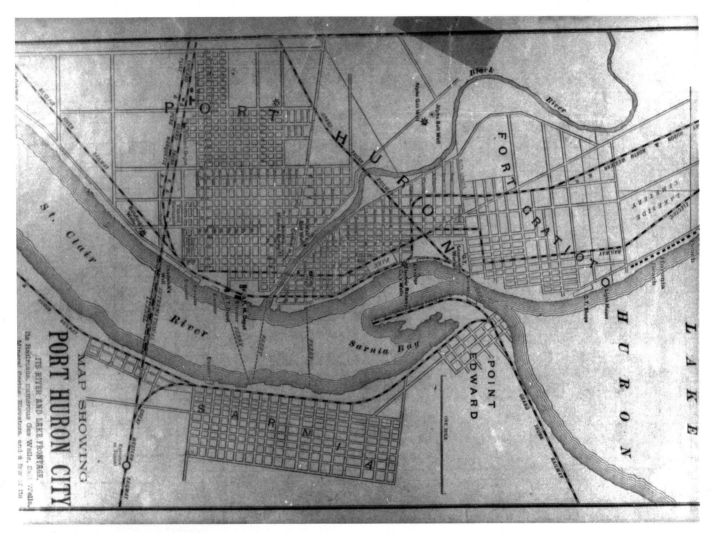

Map of Port Huron–Sarnia from the back cover of the Port Huron Times *special "Tunnel Opening" edition, September 19, 1891.*
— St. Clair County Library

CHAPTER TWO:

The Digging Begins

Hobson's first action was to verify Shanley's report. In 1885 he had a series of test borings made along the proposed route of the tunnel. Eleven holes were bored 50 feet south of the centre line, so as not to disturb the river bed on the line of the tunnel. Core samples showed the same blue clay that had foiled the tunnel attempt at Detroit in 1872.

Grand Trunk management decided to dig a test drift in this blue clay to assess the feasibility of a full-size tunnel. A contract was let to General William Sooysmith and Company of New York. Work commenced on this preliminary drift in December 1886. A shaft 14 feet by 6 feet was sunk on each side of the river, to a depth of 80 feet on the Canadian side and 90 feet on the American side. The drifts were extended, horizontally, 186 feet on the Canadian side and 20 feet on the American side before work was halted.

Sooysmith's fears were realized when natural gas erupted from the rock below, on the Canadian side, and sand and water entered the drift. After futile attempts at pumping, the drifts were allowed to fill with water and were finally abandoned in July 1887, after about nine months of work. The Grand Trunk would not relent on the terms of the contract and allow Sooysmith any compensation for the unfavourable conditions found. General Sooysmith abandoned his contract, forfeiting a $15,000 bond rather than risk losing more money. This led to a legal battle between the Grand Trunk and Sooysmith's company, the railway claiming he had failed to fulfill his contractual obligations. At one point the Grand Trunk impounded Sooysmith's equipment, but their own lawyers advised them they were overstepping their bounds.

In fairness to General Sooysmith, it should be noted that he recommended either the freezing method of excavation or the use of a shield to dig the experimental headings. The Grand Trunk was unwilling to spend the required amount on a trial shaft.

After Sooysmith's failure to complete the test drifts in the treacherous blue clay, and the subsequent lawsuit filed by the Grand Trunk, Tyler was unable to interest other contractors in bidding on the tunnel construction. Tyler then obtained the consent of the Grand Trunk directors to proceed without a contractor. The Grand Trunk management assembled their own work force. On April 20, 1888, a second attempt—this time a full-size tunnel—was started. Shafts 23 feet in diameter were dug on each side of the river and lined with brick masonry. Their idea was to avoid excavating the long approaches to the tunnel until success was assured. This plan was condemned to failure as the walls of the shaft continued to collapse. The digging finally had to be abandoned when the Canadian shaft was down within 20 feet of the tunnel level.

On May 7, 1888, a second, more detailed series of test borings was begun. This time a total of 110 samples were taken at intervals of 20 feet, exactly in line with the proposed route. When these tests were completed on July 16, they confirmed that the river bed consisted of a thin layer of blue clay. With only 10 to 12 feet above and the rock 12 to 14 feet below, this was a precariously thin layer in which to bore a tunnel 21 feet in diameter. (Workers later claimed that while digging the tunnel they could hear the bands playing aboard steamers passing overhead. One long-time railroad conductor said, "When the tunnel was new you could hear the steamers passing over and now you can hear an Indian paddling a canoe.") As well, there remained the danger of gas intruding from below. It was decided to locate the tunnel a minimum of 10 feet above the bedrock to avoid this danger. Conventional tunnelling methods would not be possible under such conditions. Shield tunnelling, a relatively new technique in tunnel-building, would be employed. Shield tunnelling had been invented in the early 1800s to excavate bores through soft earth, where standard mining techniques would not work. A shield is a steel shell that is moved through the soft earth to protect the tunnel heading from collapsing until the permanent lining can be put into place, be it masonry or, in the case of the St. Clair Tunnel, cast iron.

Tyler had become acquainted with shield tunnelling on his frequent trips to Grand Trunk shareholders meetings in London, England. Sooysmith had suggested it as one alternative for the test drifts in 1887, but the Grand Trunk had rejected the idea as too expensive for a test bore. The first tunnel built by using the shield method was excavated under the Thames River by Marc Isambard Brunel, beginning in 1818. This shield was a primitive affair 22 feet 3 inches high by 37 feet 6 inches wide. The next recorded use of shield tunnelling was in 1869, when Peter Barlow bored a small tunnel, also under the Thames River. On this conti-

nent, Alfred Beach constructed a pedestrian tunnel under Broadway in New York. About this time, a tunnel was also being bored under the Mersey River, in England, to carry an aqueduct. This tunnel was being driven through conditions similar to those confronting Hobson. Tyler returned to Canada, bringing the news to a desperate Hobson. Two attempts at driving a tunnel beneath the St. Clair River, in as many years, had failed and the situation was becoming critical for the Grand Trunk.

With the meagre information supplied by Tyler and some small drawings of Beach's shield, Hobson set about designing the shields that would be used in building the St. Clair Tunnel.

Unable to locate a U.S. manufacturer willing to build one of the shields, Hobson had them both produced at the Hamilton Bridge and Tool Company, Hamilton, Ontario. He was then required to pay customs duty on the shield used on the American side of the tunnel. The shields were shipped in sections and assembled at their respective sites. The shields, the largest of their type produced to date, were cylindrical, 21 feet 6 inches in diameter and 15 feet 3 inches long. They were fabricated from 1-inch-thick steel plates and weighed 80 tons each. They bolted together with flanges on the inside, thus presenting a smooth surface to the slippery blue clay through which they would move. The interior of the shields were divided by three vertical partitions to provide support and two horizontal partitions which also served as platforms for the workmen digging the clay. These partitions extended 11 feet back into the shield. The leading edges of the shield were knifelike to aid in cutting through the clay. A bulkhead closed off 10 of the

The Beach Hydraulic Tunnelling Shield

Detail from the Port Huron Times *special "Tunnel Opening" edition, September 19, 1891, showing various views of the St. Clair Tunnel. (Taken from* Scientific American, *August 1890.)*

– Port Huron Museum of Arts & History

The assembled Canadian shield at the top of the north slope.   – Lambton County Library

The Canadian shield being lowered into position. – Port Huron Museum of Arts & History

12 compartments created by the partitions. The other two compartments had heavy iron doors, through which all the excavated material passed. These doors could also be quickly closed if the tunnellers encountered an infusion of water or sand. Thus the heading would be protected if the need should arise.

Equally spaced around the circumference of the shield and facing rearward were 24 hydraulic rams. These were 8 inches in diameter with a stroke of 26 inches. By applying pressure to these rams, the shield could be moved forward. The rams could be worked individually, allowing the surveyors to correct for any misalignment as the tunnel progressed. They were powered by Worthington pumps which were capable of producing 5,000-pounds-per-square-inch pressure, which amounted to 125 tons per ram, or 3,000 tons total for the 24 rams. This amount of pressure was never required in the soft blue clay. The most ever used was 1,700 pounds per square inch or 40 tons per ram, for a total of 960 tons.

J.T. Eames, mechanical superintendent, employed a novel method of placing the shields in position at the tunnel portals. Rather than assemble them in place where they would begin the bore, they were erected at the north side of their respective cuts. Ramps were then constructed from four 12-inch-square timbers placed about 4 feet apart. Six large ropes were attached to the upper ends of these ramps and passed around the shields. The ropes were then coiled around pilings. As the huge shields were eased forward over the brink of the cuts, teams of workers strained at the ropes to control the descent. It took only 80 minutes from start until the shields rested upright at the bottom of the cuts, ready to begin the arduous task of boring beneath the river.

Frameworks constructed from 12-inch-square timbers were placed behind the shields. These would bear the strain of the initial load as the shields were forced into the earth ahead of them.

Tunnelling began on the American side on Thursday, July 11, 1889. A landslide on the Canadian side delayed the use of the shield until Saturday, September 21, 1889.

The following year, on Saturday, August 23, 1890, only 15 feet separated the two shields. An auger hole was bored through and a plug of chewing tobacco was passed from one side to the other. This tobacco plug has been referred to as the "first freight" to pass through the tunnel. (A *Scientific American* article of the time states that a 6-foot drift was run between the shields when they came within 125 feet of each other, and that it was lined with pine timbers. It seems highly unlikely that Hobson would take the chance of putting the entire project in jeopardy when it was so near completion. The same article contradicts itself by later stating that an auger hole was bored through on August 23 and a plug of tobacco passed from one side to the other.)

At noon on August 25, 1890, an opening was shovelled through, and Hobson, Charles Mackenzie (a director of the company), Dr. Johnston (chief medical officer), J.T. Eames (mechanical superintendent) and Thomas Murphy (superintendent of excavation) walked through. A signal was given and every whistle in Sarnia and Port Huron was blown in honour of the completion of the bore. By Saturday, August 30, 1890, at 11:30 PM. the two shields met edge to edge—exactly in line

'The meeting of the great shields of the St. Clair River railway tunnel', from Scientific American, September 13, 1890.
– Metropolitan Toronto Library

The shield begins to burrow into the earth below the St. Clair River. Sixteen rows of the cast-iron lining have been installed. The timber cribbing which the shield pushed against remains in place. Earth was removed through the hole in the top of the lining until the tunnel had advanced far enough for the lining to take the strain of the rams.

– Sarnia Historical Society/Mrs. W.L. Kirby

horizontally and out only ¼ inch vertically. The Grand Trunk had exercised great care in sighting the tunnel, as even a small error could have rendered the bore useless for rail traffic. They even went so far as to purchase a house on the American shore and have it moved out of the surveyors' line of sight.

To line the tunnel the Grand Trunk managers decided something sturdier than masonry was required to overcome the forces of the semi-liquid blue clay through which the bore passed. They chose cast iron. During his visits to London, Sir Henry Tyler had chanced to see the pedestrian tunnel which passed under the Thames River at Tower Hill. It had been constructed in 1868-69 using this method. Hobson adapted this idea to suit conditions in the St. Clair Tunnel. The tunnel lining consists of a series of rings made up of cast-iron segments bolted together. Each ring consists of 13 segments 5 feet long by 18¼ inches wide and 2 inches thick, plus one key piece 9⅞ inches long. The circumferential flanges were 2⅜ inches thick and cored with twelve 1-inch holes spaced 4½ inches apart. The radial flanges were 2¾ inches thick, tapering to 1⅝ inches at the edge. These were cored with four 1-inch holes. It required 156 bolts, ⅞ inch by 8 inches long, to bolt each ring to its neighbour. Besides the bolt holes, a 1½-inch-diameter hole was cored into each segment. This hole was used to pour cement grout between the cast-iron lining and the clay, to a thickness of 3 inches, in the lower half of the completed tunnel. In addition to the cement, on the outside of the lining, the lower half of the interior was also covered with masonry to protect it from brine dripping from passing refrigerator cars. An 1890 Sarnia *Observer* article notes that the railway had ordered two million bricks for this purpose. Between each radial joint was placed a piece of white oak 3/16 inch thick. This absorbed moisture from the clay and swelled, causing the joint to be sealed. The circumferential joints were sealed with a layer of asphalt-coated canvas roofing material. The segments were also machined at their edges with a groove that could be caulked with soft lead. Unfortunately, in the past 100 years this caulking has deteriorated, allowing water to drip through into the tunnel, causing minor problems for the railway.

The castings were poured at the Grand Trunk shops in Hamilton and at the Detroit Wheel and Foundry Company works, then delivered to the site, where they were planed to size. The ever-frugal Grand Trunk used a mixture containing 80-percent old car wheels and 20-percent Scottish pig iron. Each segment weighed between 1,000 and 1,050 pounds. A total of 56 million pounds of cast iron was used to complete the tunnel lining. After machining, the segments were heated and dipped in hot pitch to help protect them.

To install these heavy segments inside a tube 21 feet in diameter precluded the use of an ordinary derrick. Mr. T.C. Tepper, chief engineer of the Hamilton Bridge and Tool Works, manufacturers of the shields, designed a special crane for the purpose. This "segment hoist," as it was called, was attached to a centrally located shaft on the rear side of the shield. As the shield pushed forward, it exposed the area where the next cast-iron ring was to be installed. By means of three different sets of gears, the hoist could then be extended to pick up a segment, rotate it to bring it into position, and move toward or away from the shield to align the segment with the ring in which it was being installed.

The hoist grasped the segments by engaging a series of pins into some of the bolt holes cast into each segment. This held the segments secure until the bolts could be placed. Beginning at the bottom, centre, the segments were partially bolted on alternating sides until the key piece had been reached at the top. Crews working on light staging followed, completing the bolting job. It took the 15-man crew about 45 minutes to position one ring of the lining.

While provision was made for lateral alignment of the shields, nothing prevented its rotation. As the American shield moved forward it rotated towards the north until February 1890, when it was approximately 20 degrees from its original position. It then revolved in the opposite direction until it was 30 degrees to the south. The Canadian shield rotated to the south from the beginning until it was 30 degrees from horizontal. When the two shields met, they were in almost the same degree of rotation.

Once the shield was in position, the Herculean task of burrowing beneath the St. Clair River began. The job of removing the tons of blue clay that lay ahead was one of brutal, back-breaking manual labour every inch of the way. Various types of shovels and spades were tried, but none were completely satisfactory in the confined space of the shield. One of the diggers, John Ordowski of Port Huron, invented a tool based on a carpenter's drawknife and had one made by a local blacksmith. His supervisors were so impressed that in short order they had every digger equipped with one. Ordowski said later that his only compensa-

tion was that his employers and fellow workers thought he had done a good job.

As the shield was pushed forward, it cut into the clay like a giant cookie cutter. When the forward compartment had partially filled, the excavating crew gouged out the clay and passed it back. Here, it was loaded onto four-wheel carts holding 1 cubic yard. These carts were hauled by horses, up a light railway, to the tunnel mouth, where a derrick powered by a 50-horsepower Lidgerwood hoisting engine lifted the body from the car and hoisted it halfway up the slope. From there, a second derrick took it the rest of the way to the top and dumped it on a waiting flatcar. As the flatcars were filled, they were drawn, first by teams of horses and later by locomotives, to tracts of land purchased by the Grand Trunk. These tracts of land eventually became Sarnia Tunnel Yard and Tunnel Yard Port Huron. Mules replaced the horses in the tunnel when the use of compressed air began, because the horses were unable to work in the high pressure.

Where the clay was fairly stiff, the tunnellers worked 2 or 3 feet ahead of the cutting edge of the shield. But where they encountered soft earth or gravel, or where water intruded on the works, conditions demanded that they keep the shield close to the working face of the tunnel. The diggers constantly tested the earth 8 to 10 feet ahead of the shield with augers, checking for water, gas and quicksand. Occasionally large boulders were encountered and had to be broken up to be removed. Some boulders were too large to be handled this way. Holes were excavated at the side of the shield and the boulders were rolled out of the way.

Preparations are being made for the masonry work at Sarnia Portal, 1890.   – Gene Buell Collection

Masonry on the Sarnia Portal is almost complete in this photograph, but the approach has yet to be excavated.   – Gene Buell Collection

In some instances the ground was so soft that the shield was forced ahead far enough to completely fill the forward section, with earth pouring through the bulkhead doors, where it was scooped up. The average monthly rate of advance in the tunnel was 236.5 feet on the American heading and 219 feet on the Canadian side. The following table shows the monthly progress on both headings in feet.

	Canadian Heading	American Heading
1889		
July		53.00
August		144.05
September	73.30	153.70
October	109.45	126.75
November	187.50	225.50
December	217.40	266.91
1890		
January	292.35	277.59
February	306.08	273.67
March	292.50	203.63
April	281.34	182.20
May	97.00	355.54
June	236.83	354.46
July	201.30	382.30
August	334.04	311.11

An infusion of water and gas had halted the trial shafts and headings that Sooysmith had attempted, and the Grand Trunk was ever fearful another such occurrence would condemn the whole project to failure. A Davy safety lamp was used to monitor the air in the shaft for gas. Ventilation was provided by two Roots blowers on each heading. These had a capacity of 10,000 cubic feet per minute each. Should gas be struck, this reduced the risk of explosion to almost nothing.

Adequate ventilation solved the problem of gas explosion in the tunnel, but the intrusion of water still posed a threat to the completion of the project. It was feared that the bottom of the river might collapse into the workings, inundating them. When the shields reached the river banks, a point 1,716 feet on the Canadian side, 1,994 feet on the American side from the commencement of shield tunnelling, Hobson instituted the use of compressed air in the tunnel. Hobson later stated that he would use compressed air sooner on a second tunnel because it reduced the amount of material which crept into the excavation.

Brick bulkheads, 8 feet thick, were constructed near the mouth of each shaft. Into these were built two air locks 7 feet in diameter and 16 feet long, and one 12 inches in diameter and 25 feet long. These air locks had heavy iron doors at either end. After men and materials entered, the door was shut and appropriate valves were operated to allow the air pressure inside the chamber to equalize with that in the section of tunnel to which the worker was going. When the pressure had equalized, the door on the other end of the chamber could be opened and the passage into or out of the tunnel could be completed. Everything that went into or out of the tunnel passed through these air locks. The men and horse-drawn cars carrying clay used the two larger ones. The smaller one was used for lengths of pipe too long to go through the larger locks.

Pressurized air was supplied, on each side of the river, by two 20-by-24-inch compressors made by the Ingersoll Seargent Rock Drill Company of New

York. They were capable of delivering 570 cubic feet per minute. Unlike modern shield-tunnelling projects, where only the diggers work in compressed air, Hobson's shields were not made airtight, and thus the entire tunnel had to be kept pressurized. This meant that everyone working in the tunnel, not just the excavators at the face of the bore, had to work under pressures which sometimes were as high as 28 pounds above atmospheric. Many of the men suffered from what they termed "tunnel grip" because of working in the pressure. Three men, Daniel Thayer, George W. Culley and John Johnsick, died from caisson disease, more commonly known as "the bends," an extremely painful condition caused by too-rapid decompression.

In March 1890 the men digging at the Canadian heading sent a round-robin letter to Thomas Murphy, superintendent of excavation, complaining of working conditions. They felt that working in the pressurized air was hard on them physically. (*Harper's Weekly* for February 28, 1891, reported that the use of compressed air started on April 7, 1890, on the U.S. side and May 20, 1890, on the Canadian side. The article also stated that compressed air was used until October 2, 1890. These dates disagree with the letter of complaint which was received by Murphy on the afternoon of March 8, 1890, fully a month before *Harper's* says compressed air was used in the tunnel. The finish date is possibly correct, as the shields met on August 30, 1890, but it would take time to finish excavating and put the lining in place, and Hobson obviously would not have taken any chances at that time by stopping the compressors too soon.)

The men asked for a pay raise, from the 17½ cents per hour they were receiving to 25 cents per hour. By the time the company responded, the shields had met and the work under pressurized conditions had been completed. The company cut the workers' pay to 12½ cents per hour but allowed them to work 14 hours a day, instead of 8, to make up the difference in their take-home pay!

Despite the complaints of the workers, conditions in the tunnel were better than on most contemporary excavating jobs. Electric lighting and adequate ventilation made working conditions relatively good.

As the shields approached the completion of the bore, contracts were let for other portions of the project. On August 18, 1890, Nihan, Elliott & Phin of St. Catharines, Ontario, contracted to excavate the long approaches to the tunnel. These were 3,100 feet on the Canadian side and 2,500 on the U.S. side. It is interesting to note that this contract was let only 12 days before the shields actually met on August 30, 1890—perhaps a sign of the GTR's uncertainty about the successful completion of the tunnel in such soft clay.

The contract for the masonry portals and retaining walls was let to Wm. Gibson, MP, of Beamsville, Ontario.

Several landslides delayed the work, requiring the widening of the cuts and enlarging of the retaining walls. This added to the costs and delayed the opening of the tunnel. During the last five weeks before the opening ceremonies, work was carried out around the clock to ensure completion on time.

After completion of the actual digging in the bore, much remained to be done. The interior partitions were removed from the shields and the lin-

The round-robin letter complaining about working conditions. (The reason the signatures are in a circle is so that their supervisors couldn't tell who the instigators of the complaint were.) — CNR

St Clair Tunnel Co.

Sirs we the persons whose signatures encircle this requisition and who represent the three gangs which dig in front of the shield on the Canada side, being your employes do hereby request that our wages be raised to twenty five cents per hour, and eight hours a days work as usual Our reasons for asking for this advance are 1st That our work is under the ground away from the sunlight and natural atmosphere 2nd The work is much more laborous than any ordinary labour this and the closeness of the air around us makes our days work a continual succession of heats 3rd It is dangerous work in many ways The effect of these disadvantages we find to be very hard on our physical constitutions breaking down many and making it necessary for all to lay off frequently to rest up and recuperate our health. Signed yours Obt Servants

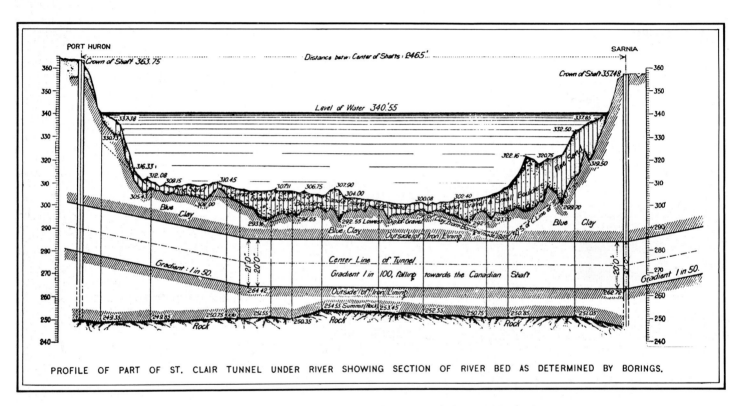

PROFILE OF PART OF ST. CLAIR TUNNEL UNDER RIVER SHOWING SECTION OF RIVER BED AS DETERMINED BY BORINGS.

From Engineering News, October 4, 1890. (the gradient falling towards the Canadian side is incorrectly shown as 1% instead of .1%. The scales indicate height in feet above sea level.)

Excavation of the approaches began after the tunnel bore had been completed.
– Lambton County Library

ing was assembled through the shields. The shells of the shields remained exactly where they had met. This was done to eliminate the possibility of collapse of the last few feet of the soft blue clay, which would have condemned the entire project to failure. Since the heavy segments (approximately 1,000-1,050 pounds each) which make up the tunnel lining were lifted into place by the special crane attached to the rear of the shields, it is interesting to speculate how the last segments were installed after the interiors of the shields had been removed.

The brick bulkheads and iron doors which made up the air locks were removed. The clay roadbed and the temporary tracks of the horse-powered rail lines were removed. The tunnel was cleaned and concrete was forced through the holes in each segment into the space between the earth and the bottom half of the lining. Brick masonry was then applied to the lower part of the tunnel interior to prevent damage from brine dripping from passing refrigerator cars.

The permanent roadbed was installed and rail weighing 100 pounds per yard, the heaviest used on the GTR, was laid. The St. Clair Tunnel had cost approximately $2,700,000, of which $375,000 was a subsidy from the Canadian government.

Expenses were broken down as follows:

Expended on preliminary work	$250,000
Machinery and plant	$250,000
Labour	$900,000
Cast iron for lining	$800,000
Other materials	$100,000
Real estate, land damages, legal expenses, etc.	$110,000
Permanent expenses (track, locomotives, etc.)	$50,000
Approaches	$200,000
Engineering, superintendence, etc.	$40,000
Total	$2,700,000

Tyler had considered, but rejected, the idea of laying a giant tube across the river. Some 20 years later, the Michigan Central did just that at Detroit.

On April 9, 1891, history was made when Sarnia yard engine GTR 4-4-0 #253, in the charge of engineer Wm. Ovens, fireman Wm. McNeish and conductor Nelson McKee, made the first trip through the tunnel carrying Hobson, assistant general manager Charles Percy of Montreal, Col. Tisdale of Simcoe, and a number of Grand Trunk employees. The train proceeded from the Canadian side to the American side, remained a short while and returned. Steam whistles sounded on both sides of the river.

The handwritten caption on the back of this original photograph says, "First crew through the tunnel."
— Irene McLean

GTR *4-4-0 #253 made the first trip through the tunnel on April 9, 1891.* – Mike Smout Collection

Dignitaries wait to board the inaugural train for the official first trip through the tunnel, September 19, 1891.

– Douglas Wilson

CHAPTER THREE:

The Steam Years

On Friday, September 18, 1891, dignitaries and officials of the Grand Trunk arrived at Sarnia on every train. Preparations for a celebration had been under way all week. Sarnia, Port Huron and Point Edward were a frenzy of activity. The GTR was finally going to have its day. The tunnel was completed; only minor details needed attention before revenue traffic could commence. Floral arches had been erected at both ends of the tunnel. The Sarnia freight house had been cleared of goods and decorated in anticipation of Saturday night's banquet.

Ceremonies began Friday evening when the citizens of Port Huron presented a banquet in honour of Sir Henry Tyler and the GTR. Sir Henry and the other guests of honour departed Point Edward aboard the ferry *Omar D. Conger*, landing at Huron Street at 7 PM., from whence they proceeded to the Huron House for the festivities. Over 100 people attended, among them, L.J. Seargeant, general manager of the GTR; E.P. Hannaford, chief engineer of the GTR; Joseph Hobson, chief engineer of the tunnel; W.J. Spicer, manager of the Chicago & Grand Trunk Railway (it should be noted that the C&GT was operated as a subsidiary of the GTR of Canada until 1898); Michigan's Governor Winans; Congressman Justin R. Whiting; Mayor Watson; and the Honourable Charles Mackenzie of Sarnia.

Mayor McIlwain of Port Huron presided at the banquet. In his address, he congratulated Sir Henry Tyler and the GTR on the completion of the tunnel. During the evening Sir Henry was pre-sented with a silk sachet bag made from the first American flag to pass through the tunnel. It was made by Mrs. Alice Mann from a flag which she had carried through the tunnel herself. The guests retired early and returned to Point Edward to prepare for the big day ahead.

The formal opening of the tunnel occurred on Saturday, September 19, 1891. Sir Henry Tyler and other GTR officials boarded a train at Point Edward depot for the short trip to the tunnel. They departed at 12:30 PM. The sky was cloudless and the temperature was 62 degrees. A brief stop was made at Sarnia Station to take on more guests from Sarnia and Port Huron. They then proceeded, arriving at the summit of the tunnel approach a few minutes later. The official train consisted of Chicago & Grand Trunk 4-4-0 #148, built by Rogers Locomotive Works in 1891, and seven open-vestibule Grand Trunk coaches in the charge of engineer Whittaker and conductor Shaw. The *Erie Limited Express* arrived, carrying a number of distinguished guests from points as far away as New York City.

The official opening did not coincide with the actual commencement of revenue traffic. Sir Henry was leaving for England and the official ceremonies had been moved up for his convenience.

When the official party had detrained, Mayor Watson of Sarnia presented an address by the town council, congratulating Sir Henry Tyler and the GTR on the successful completion of the St. Clair Tunnel. Sir Henry responded briefly, the band of the 27th Battalion played "God Save the Queen," and the official party entrained. At 1:20 PM. the train moved forward through the floral arch, down the incline toward the east portal, and into history. Throngs of cheering spectators lined the banks as the train disappeared into the tunnel. Three and one-half minutes later, it emerged on the American shore to still more cheering crowds. Ascending the grade, it passed through the second floral arch and proceeded to the new depot at 22nd Street. Near the summit stood one of the special 0-10-0T engines built for use in the tunnel.

Sir Henry and his party disembarked at the U.S. Customs House, which had been fitted with a platform for the dignitaries and decorated with evergreen boughs. Mayor McIlwain welcomed the group, adding his congratulations to the others already bestowed upon Sir Henry, the GTR and the St. Clair Tunnel Company.

A note of humour was injected into the ceremony by Sir Henry when he told the audience that, at the banquet the preceding evening, he had been given an American flag to carry through the tunnel on the inaugural train. He then reached deep into his coat pocket, pulled out what he thought was the flag, and waved it joyfully above his head. It was not until the crowd roared with laughter that Sir Henry noticed he had taken a white handkerchief from his pocket. Sir Henry quickly tried another pocket, only to find a second white handkerchief. The crowd cheered and roared as Sir Henry made a futile search of his pockets for the elusive flag. Finally Mr. Seargeant, general manager of the GTR, came to Sir Henry's rescue by presenting him with the missing flag, stating that his prize was so greatly coveted that

they had actually stolen it out of his pocket on the way over. Sir Henry concluded his remarks and the band of the 13th Battalion, Hamilton, played the "Star Spangled Banner" and "Hail Columbia." The official guests boarded the train and departed for the C&GT depot at the foot of Griswold Street. From there, they took the ferry *Omar D. Conger* back to the Sarnia freight dock to participate in the feast which had been prepared in the GTR freight shed. The freight house had been decorated for the banquet by Mr. Atchison, chief of the Hamilton Fire Brigade, assisted by a Mr. Anderson, also of Hamilton. The meal was provided by Montreal caterers. The wines alone were reported to have cost $4,500. The guest list read like a who's who of politics, industry and railways in Canada and the U.S., and included such notables as Sir Casimir Gzowski; Governor Winans of Michigan; George Pullman, inventor of the Pullman Car; Sir John Ross, Commander of Her Majesty's Forces in Canada; the mayors of Sarnia and Port Huron, as well as those of Montreal and Detroit; W.J. Spicer, manager of the Chicago & Grand Trunk Railway; Joseph Hobson, chief engineer of the tunnel; L.J. Seargeant, general manager of the GTR; and E.P. Hannaford, chief engineer of the GTR.

Toasts were made to the Queen and to the President of the United States. Sir Henry then presented the audience with a history of the St. Clair Tunnel Company. He praised the men in charge of the project, particularly Joseph Hobson, who had designed and supervised the use of all the special equipment required in the construction of the tunnel. He then called upon Hobson to give a speech.

The audience gave Hobson a standing ovation, cheering and waving napkins. Hobson responded

Grand Trunk Railway
OF CANADA

Admit *P. H. Phillips Esq.,*

TO THE

INAUGURAL TUNNEL BANQUET

AT

SARNIA, SATURDAY, 19TH SEPTEMBER, 1891,
AT 2 P.M.

THIS CARD WILL ALSO PASS THE BEARER BY THE
INAUGURAL TRAIN THROUGH THE TUNNEL.

L. J. Seargeant
Genl. Manager

Floral arch at Sarnia on September 19, 1891.

– Lambton County Library

graciously, thanking Sir Henry for his unwavering support and determination in completing the tunnel. He paid tribute to Messrs. Hickson and Seargeant, the managers to whom he was directly responsible: "No engineer ever had less cause to complain of annoyance or interference from those above him." Hobson had been allowed to choose his own assistants. He spoke of Mr. Hillman's accuracy in sighting the progress of the tunnel so that the shields met only one quarter of an inch out of line in 6,000 feet. He praised Thomas Murphy, who had charge of the excavation, for instilling confidence in his workers. He commended J.T. Eames, who was in charge of machinery, for keeping everything operating smoothly. The banqueters adjourned from their celebrations in good spirits.

With the formal opening ceremonies over, the St. Clair Tunnel Company put the finishing touches on the yards and structures. The first revenue freight traffic moved through the tunnel on Saturday, October 24, 1891, when a train of 11 cars travelled from Port Huron to Sarnia behind engine #598, which had been specially decorated for the occasion with the Union Jack, the Stars and Stripes, and coats of arms of both countries emblazoned on her sides. (A 1925 article in the Port Huron *Times-Herald* recalls the memories of one Frederick F. Minard, the engineer on the first freight train. He lists the date as 29 September, 1891 and states the train had 16 cars. This differs from contemporary accounts.)

Four special locomotives were built for operation in the tunnel. These engines, the largest in the world at that time, were of the 0-10-0T wheel arrangement and were referred to as "Decapod" type by the Grand Trunk.

Even though she was the newer of the two, the car ferry *Huron* was laid up immediately and other jobs found for her crew. The *International* continued to ferry passenger trains at the Point Edward–Fort Gratiot crossing until the depots at Port Huron and Sarnia Tunnel were readied to accept passengers.

The first passenger train passed through the tunnel on December 7, 1891. The ferry service was then discontinued and the *International* was laid up with the *Huron* at Sarnia. They remained idle until 1898, when the *International* was sold to Hiram Walker's Lake Erie & Detroit River Railroad and the *Huron* was transferred to the Grand Trunk's ferry operation at Windsor. The ferry docks at Point Edward were later converted to ore unloading docks by the Grand Trunk, to facilitate movement of iron ore from Great Lakes' boats to Stelco's plant in Hamilton. The footings of these docks can still be seen near the base of the Bluewater Bridge.

With the opening of the St. Clair Tunnel and discontinuance of the ferries at Point Edward, traffic patterns on the Grand Trunk changed. Through passenger traffic between Chicago and the East was routed by way of the Southern Division through London, rather than the Northern Division through Stratford to Toronto. The depot at Point Edward was closed and eventually removed, and a new line around the east side of Sarnia connected the Northern Division main line with Sarnia Tunnel Station. Freight traffic only was moved to Point Edward, and the old Great Western depot on Sarnia's waterfront was used mainly

All four 0-10-0T's shortly after delivery from Baldwin. — PAC

Port Huron depot was completed in February 1892. It twice suffered fire damage to its interior, in 1893 and 1908, but was repaired both times. It was finally demolished in 1975.

– St. Clair County Library

for transshipment to boats for shipping up the Great Lakes. The tunnel cut approximately two hours off the time between Chicago and Toronto, and saved the Grand Trunk $50,000 a year in operating costs of the ferries. It also eased the situation with regard to the backlog of cars awaiting transfer across the river on the ferries.

The tunnel, however, was not without its problems. This appeared in the form of gases produced by the large steam locomotives used to haul trains through the tunnel. These engines had been designed to burn coke or anthracite coal in order to reduce the amount of smoke produced. It still required 45 minutes for the ventilation fans to clear the gases completely from the tunnel, although it was considered safe to enter the tunnel after 15 minutes.

On Sunday, January 31, 1892, only three months after revenue service had begun, the first serious accident occurred when an eastbound freight train broke in two in the tunnel. The engineer proceeded out of the tunnel with the cars still attached to the engine. Conductor George Hawthorne of London and brakeman Joseph Whalen of Point Edward climbed down from the caboose, but they were soon overcome by the fumes. They were found, but efforts to revive Hawthorne failed, and he thus became the first victim of asphyxiation in the tunnel.

A report in the Sarnia *Observer* in February 1892 noted that the railway was having difficulty getting men to work in the tunnel despite the fact brakemen were earning $2.50 a day, a quite substantial wage at that time. The problem of the fumes was obviously well known.

The problem of trains parting in the tunnel was also very common. The old-style link-and-pin couplers then in use were susceptible to breakage. As the train rolled down the grade toward the centre of the tunnel, the cars would bunch up. As the engine began its ascent out of the tunnel, the train would run out, putting a strain on the couplers, and if the engineer had even a little bit of a heavy hand on the throttle, he could snap the train in two. This problem was somewhat remedied with the introduction of both air brakes and knuckle couplers in the mid-1890s. It took a few years for all cars to be converted, however, and on Sunday, November 28, 1897, a second fatal accident occurred.

This time train #34 eastbound departed Port Huron about 8:30 P.M. with 26 cars of provisions, meat and vegetables, plus a caboose, all hauled by engine #599. It parted approximately 200 feet inside the Canadian portal of the tunnel. Engineer P.J. Courtney returned to pick up the lost cars. Stationed at the Canadian pumphouse, located at the mouth of the tunnel, duty engineer John Tonkin heard the engine when the train parted. Since this was a fairly common occurrence, he didn't worry until the train failed to reappear. He then notified the yardmaster, Frank McKee. McKee, Tonkin and car-sealer J. Lowe entered the tunnel on foot. They found brakeman Wm. Potter unconscious near the coupler that had parted. Fireman Wm. Duncan was also unconscious on the ground. They placed these men on the tender and climbed into the cab, only to find engineer Courtney dead and the engine in reverse. Yardmaster McKee proceeded to take the engine out of the tunnel. Doctors who had arrived on the scene revived Duncan and Potter. Conductor Arthur Dunn and brakeman John Dalton, who had been in the caboose, were also overcome by fumes and died before they could be taken out of the tunnel.

At the inquest, chief engineer Joseph Hobson stated that the Grand Trunk was looking into the possibility of using electric power in the tunnel—something they had considered after the first asphyxiation in 1892.

Seven years passed before the tunnel claimed its next victims. On Sunday, October 9, 1904, an eastbound freight train consisting of 16 cars of coal, hauled by engine #1301, parted in the tunnel. With the 1897 tragedy still fresh in their minds, engineer John Coleman and fireman Fred Forester took the six cars still attached to the engine out of the tunnel and returned to recouple the train. Broken couplings were quite commonplace in the St. Clair Tunnel operations, and recoupling a train in the tunnel was usually a routine job. Because of the broken coupler, the crew had to connect the cars to the engine with a chain. They uncoupled three cars, feeling this was all the chain could handle, and they removed them from the tunnel. They returned a third fateful time to recover the rest of their train, but by now the smoke and fumes were working their deadly deed. Engineer Coleman collapsed in the cab of #1301. About this time brakeman Alfred Short climbed down from the caboose to walk to the American portal. By the time he exited the tunnel, he was severely affected by the gas. He gave the alarm. Alexander S. Begg, superintendent of terminals, proceeded into the tunnel. This was the last time he was seen alive. When the rescue party from Port Huron entered the tunnel, they found his body laying on the tracks.

When the train failed to reappear shortly after its third trip into the tunnel, the men in the Sarnia yard realized something had gone wrong. Conductors Charles Fisher and Richard Tinsley, and brakemen James Hamilton, Walter Hawn, Alex Forbes, Daniel Gillies and Thomas McGrath, entered the tunnel in an attempt to rescue the train crew. Tinsley and McGrath were overcome by the fumes, and the rest barely made it out of the tunnel. A second team, consisting of conductor Porter, yardmasters Franklin McKee, John Blake and John Arbaugh, entered the tunnel on foot. They found engineer Coleman dead and the engine in reverse. Coleman was slumped against the boiler and badly burned. The fireman was found lying unconscious with his head in the water tank of the tender. Breathing the relatively fresh air in the tank had saved his life. The bodies of brakemen Daniel Gillies and Thomas McGrath were found lying beside the engine, where they had fallen. (Gillies' brother, John, had been an engineer with the Grand Trunk. He had been killed on the freight train involved in the ill-fated Wanstead wreck just two years earlier, on December 26, 1902.) Gillies and McGrath were placed on the engine, and the rescuers managed to start the engine and remove the train from the tunnel. When they emerged, they found the bodies of conductor Joseph B. Simpson and Richard Tinsley in the van.

In the summer of 1905 ten Grand Trunk employees were awarded medals by the Royal Canadian Humane Association "for conspicuous bravery in the rescue of those in danger of asphyxiation by gas in the St. Clair Tunnel on 9th October,

The Atlantic Express *exits the tunnel eastbound behind 0-10-0 #598, circa 1895. Note that a tender has been added, but she still retains the side tanks, wooden cab, box headlight and link-and-pin coupler. Notice the engineer opening the cab door to let in some fresh air.*

— NAC

1904." The list of recipients included Franklin J. McKee (who had also been involved in the rescue of 1897), John Blake, Charles Fisher, James Hamilton, Fred Forester, Walter Hawn, Alex Forbes, John Arbaugh, Wm. Cameron and Angus McDonald.

Some sources say the vans were run directly behind the locomotives as a safety precaution. This obviously was not the case in the three accidents involving asphyxiations. Air brakes were not used in the tunnel, so if the train broke in two the locomotive would not be stalled inside the tunnel spewing smoke and fumes.

Over the years stories have surfaced claiming that the trains involved in the asphyxiation accidents carried livestock and that all the animals were killed, too. There may have been a train of livestock that was overcome by fumes, but no evidence of it has been found.

While uncouplings and stalled trains in the tunnel were quite common during steam days, a strange coincidence concerning the fatal accidents in the tunnel is that all three occurred on a Sunday and they all involved eastbound trains.

By now it had become obvious to the Grand Trunk management that changes were necessary. Due to the deadly gases produced by the steam locomotives, ten men had lost their lives in the tunnel in only 13 years of operation.

Postscript: Two years after the opening of the tunnel, the Port Huron depot was engulfed by fire. Fire struck again in December 1908 and the interior of the station was destroyed. Both times the building was repaired.

Confusion exists as to the actual crew of the train involved in the 1904 asphyxiation. Most accounts agree on the names of the engineer, fireman and conductor; however, the names of brakemen are listed variously as Alfred Short, Charles Cable, John Weston and Daniel Gillies. It is believed that Gillies was the head-end brakeman riding the engine. Short was likely a rear-end brakeman riding the van, because he apparently walked out of the American portal. The whereabouts of Cable and Weston during the incident is not known, as their names do not appear among the list of injured. Contemporary newspaper accounts varied not only from one paper to another, but even in one story in the same newspaper!

#1301 prepares to couple to a westbound passenger train in front of Sarnia Tunnel Depot, circa 1906.
— Author's Collection

Sarnia waterfront during the teens. The Northern Navigation Company dock replaced the former Great Western Station in 1909. — Lambton County Library

Locomotive #1304, running in reverse with a westbound freight, crests the summit at Port Huron, circa 1908.

– John Hus Collection

0-10-0 #1303 at Fort Erie, Ontario. The engine has had its cab moved to the rear of the boiler and has been relettered "Grand Trunk."
 — Fort Erie Railway Museum

At least two steamers (1301 & 1303) had their cabs moved to the rear after the electrics came into operation.
– Port Huron Museum of Arts & History

"The Old and The New" postcard by L. Pesha. Pesha was a Marine City, Michigan, photographer who documented much of the Sarnia–Port Huron area shortly after the turn of the century.

– Bob Gray Collection

CHAPTER FOUR:

Electrification

The use of electricity to power railways was still in its infancy when the St. Clair Tunnel was conceived and built. Electric power had been applied to light trolley and street railway lines, but the technology to build and operate a heavy freight line such as the St. Clair Tunnel did not yet exist. It may seem foolish to us today that the Grand Trunk management would choose to use steam locomotives, with their deadly smoke and gases, in such a confined area as the St. Clair Tunnel. But they really had no other choice of motive power. By the turn of the century, however, heavy electric traction technology had developed to the point where moving large tonnages was feasible. This fact, combined with the recent asphyxiations, prompted the Grand Trunk to investigate the new source of power for trains in the tunnel. Bids were entertained from a number of different suppliers. The two chief competitors for the contract were General Electric, who proposed an outside third-rail DC system of 600 volts, and the Westinghouse Company, whose proposed 3,300-volt AC single-phase system was eventually chosen. The Grand Trunk specifications stipulated that the new locomotives be capable of pulling a 1,000-ton train up a 2-percent grade in 15 minutes at not less than 10 miles per hour. This was approximately three times greater than the actual requirements along the tunnel line at the time. The GTR also required that the system be able to supply power to run the pumps located at each portal, as well as electricity for lighting Port Huron and Sarnia Tunnel Stations, Sarnia roundhouse and some yard lighting.

In 1905 a contract was let to the Westinghouse Electric and Manufacturing Company of Pittsburgh, Pennsylvania, to supply the locomotives and plant. The firm of Bion J. Arnold of Chicago, Illinois, acted as consulting engineers for the railway, designing the installation of equipment.

Electricity was supplied by steam-driven generators housed in a special building constructed on the American bank of the river. Equipment consisted of four 400-horsepower Babcock & Wilcox water-tube boilers arranged in banks of two, each boiler having three drums 42 inches in diameter and 23 feet 4 inches long. The boilers were fed by automatic stokers. Separately fired superheaters were attached, and normal operating pressure was 200 pounds per square inch.

The arrangement allowed quick steaming. This was necessary because unlike in steam locomotive operation, where the fireman and engineer are in immediate contact with the situation at hand and can foresee a need for more steam, the stationary engineers in the boiler house had no direct contact with the train. By monitoring the gauges on the switchboard, they could follow the routine progress of a train from terminal to terminal, but they could not foresee any unexpected need for more power. Therefore, they needed the capability to generate plenty of steam before they lost pressure and shut down the whole system.

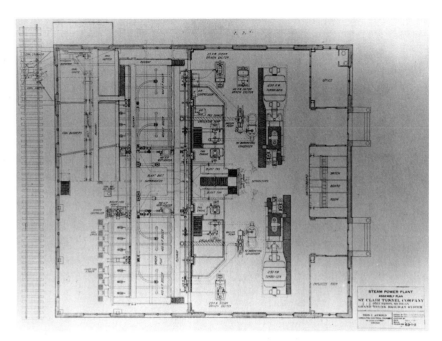

Plan of the powerhouse.
– St. Clair County Library

Powerhouse looking north. Coal was brought in for the boilers by hopper cars on a track located between the building and the water's edge.
– Tom Gaffney Collection

The generators were two Westinghouse Parsons turbo-generators designed to operate at 3,300 volts, with a frequency of 25 cycles per second. They were approximately 37 feet long, 6 feet wide and 8 feet high.

All of this fed through a large switchboard where the proper voltages could be distributed to the various points in the system, and recording gauges made logs of power consumed. The power was delivered to the locomotives by overhead catenary supported on steel towers erected 250 feet apart. The overhead system consisted of a 4/0 hard-drawn copper wire suspended by a 5/8-inch-thick galvanized-steel messenger cable which kept the copper wire at a height of 22 feet above the rail in the yards. In the tunnel, steel brackets which held the insulators were bolted directly to the tunnel lining. This supported the wire only a few inches below the top of the tunnel. Serious doubts were expressed at the time about running a bare wire carrying 3,300 volts so close to the cast-iron lining, especially since the interior of the tunnel is usually damp, but the company had no major problems because of this. The wires inside the tunnel were placed by lowering a flatcar (rigged to hold the wire spool) down the grade, using the hand brakes to control the descent. This was followed by a boxcar, which the workmen used as a platform. All the installation work was done without interruption to the traffic through the tunnel.

The six new locomotives were a joint effort of the Westinghouse Company and Baldwin Locomotive Works, who had built the 0-10-0 steamers used in the tunnel. They were of a box-cab design and rode on a rigid six-wheel frame. Although numbered individually (1305-1310), the company considered each locomotive as a "half unit," with two "half units" making up one complete locomotive. The company normally operated the electrics in pairs. Occasionally a single unit might be used on a passenger train. In later years, as train weights increased, multiples of four units became common. Each unit was bi-directional and equipped with multiple-unit controls, so any number of engines could be operated together by one engineer. This feature more than anything extended their useful life to 50 years.

The first trip under electric power was made on Thursday, February 20, 1908, when engines #1308 and #1309 ran light from Sarnia to Port Huron. There they picked up a 700-ton freight train and returned through the tunnel. The train was in the charge of conductor Walter Hawn, with superintendent Jones at the throttle. H.H. Rushbridge of the Westinghouse Company supervised the move. Over the next three months, Tunnel Company personnel were instructed in the use of the new equipment. Mixed service with the electrics and steamers was not possible, since it took upwards of 20 minutes to dry off the insulators in the tunnel after a steamer had passed through. Therefore, the electrics operated 18 hours a day and the steamers the other six.

By May 17, 1908, the men operating the electric locomotives had become familiar enough with them that the company decided to eliminate the use of steam in the tunnel completely. The steamers, once the largest engines in the world, were downgraded to yard duty in Sarnia and Port Huron. By 1920 the last of them would be cut up for scrap at Hamilton. With the departure of the steamers from the tunnel went the noxious fumes and smoke which were so hard on personnel and equipment. Steam would no longer be used in the tunnel proper. Grand Trunk trains would cut off their road power at Sarnia Tunnel or Port Huron depots, and the electrics would haul the train through the tunnel. Steam locomotives would then couple to the train at the opposite side for continuation of the journey. This procedure would be followed until the abandonment of the electrics in 1958.

The Westinghouse Company was responsible for operating the new system until November 12, 1908. On that date it was officially handed over to the Tunnel Company in formal ceremonies. Officials of the Grand Trunk and Westinghouse, along with invited guests, boarded a train at Port Huron. The train consisted of two electric locomotives and a number of flatcars fixed with railings and passenger-car seats. They proceeded through the tunnel to Sarnia Tunnel depot, where a special train took them to a reception at the Hotel Vendome in downtown Sarnia. Numerous toasts were offered to Joseph Hobson in recognition of his work in the construction of the tunnel almost two decades earlier. After the reception, the party returned to Port Huron, where they inspected the new powerhouse. In the evening they were entertained at the Harrington Hotel in Port Huron. The festivities on this occasion were not nearly as elaborate as those at the official opening ceremonies in 1891.

Train movements through the tunnel were controlled by a block signal system. The dispatcher was housed in the tower at East Summit, located at the top of the grade on the Canadian end of the tunnel. A second signal tower was located at West Summit in Port Huron. The dispatcher had absolute control over trains in the tunnel. He co-ordinated moves with the yardmasters in Sarnia and Port Huron by a yard telephone. There was also a special telephone line connecting East Summit with the powerhouse, West Summit tower, the roundhouse, and a telephone located at the middle of the tunnel. Actual dispatching of trains through the tunnel was done by telegraph from East Summit. Any problems that arose, whether with a locomotive, at the powerhouse or other places, were reported to the dispatcher immediately. He then took the necessary steps to remedy the situation. Rather than the usual written train orders commonly used in block dispatching systems, the St. Clair Tunnel Company employed a unique "staff" system to prevent having two trains in the tunnel at the same time. In each summit tower was a specially designed machine that held a supply of "staffs." These staffs were cylindrical metal rods approximately 6 inches long by ½ inch in diameter, slotted in such a way that a plain rod would not fit into the machines in the two towers, which were interconnected. It took simultaneous action by the two operators to remove a rod. Once a staff had been removed, the machines were automatically locked so that a second staff could not be removed. As the train passed the summit tower, the operator handed the staff up to the conductor, who placed it in a special receptacle at the engineer's position in the cab, where the rules stated it must remain for the duration of the trip through the tunnel. No crew would take a train into the tunnel without this staff. They also would not surrender it to the operator at the exit of the tunnel unless the train was completely

Electrics #1307 and #1308 head an eastbound train of refrigerator cars. – CNR

Electric #9152 emerging from the tunnel on the Canadian side in the late 1920s. — CNR

through, so that, for example, if the train had parted in the tunnel and they had to return to recover the lost cars, it was impossible for a second train to be in the tunnel at the same time. To facilitate handling such a small object, a special holder was used, consisting of a piece of rubber hose about 12 inches long and sealed at one end.

The system was so faultless in its simplicity that not one accident attributable to dispatching ever occurred in the tunnel. Today this is all gone, the summit towers have been razed and the staff system has been replaced by Centralized Traffic Control, or C.T.C.

A couple of interesting events incidental to the actual operation of the tunnel occurred shortly before electrification of the tunnel. An Italian immigrant woman who spoke no English was travelling to Chicago to join her husband, who had arrived shortly before. The woman, who was very much pregnant, began to experience labour pains as the train entered the tunnel. The conductor called for a physician among the passengers, but none was available. A veterinary surgeon offered his assistance and delivered a healthy baby girl. The woman refused to disembark at Port Huron and continued her journey to Chicago with her new offspring. The passengers passed the hat and collected about $200, which was presented to the newborn infant.

The second event, of a more ominous nature, happened in 1917, at the height of World War I. Albert Karl Kaltschmidt, president of Marine City Salt Company, and several other conspirators were convicted of attempting to blow up the St. Clair Tunnel. Evidence introduced at the trial indicated that Kaltschmidt was the head of a German secret

service in Michigan, whose purpose was to bomb important industrial targets in southeastern Michigan and southwestern Ontario. They planned to destroy the tunnel with a charge of dynamite that was to ride the rails into the tunnel on a platform attached to roller skates. Needless to say, their plan failed, but had it succeeded, it would have had serious consequences for the Grand Trunk. The main link which carried approximately 90-percent of the Grand Trunk's east-west traffic would have been severed, and the railway would have had to rely on the car-ferry crossing at Windsor–Detroit. Kaltschmidt received four years in Leavenworth Penitentiary and a $20,000 fine. His sister received a three-year term in the Detroit House of Correction and a $15,000 fine. Her husband, Fritz Neef, and a Carl Schmidt of Detroit each received two years at Leavenworth and $10,000 fines. Mrs. Schmidt was sentenced to two years in the House of Correction and fined $10,000. Others in the group were already serving time for bomb attacks on other area targets.

A suspected sabotage attempt occurred in 1940, when upon exiting the tunnel, a boxcar loaded with aircraft engines was found to be on fire. Arsonists had drilled holes in the floor and inserted rags soaked in flammable liquid. Sarnia Fire Department doused the flames quickly and no serious damage was done. Nothing was ever proven, but authorities suspected the arson attempt may have been aimed at a carload of high explosives which had passed through the tunnel just previously.

These attempts to destroy the tunnel indicate that it was regarded as a very important link in the North American transportation system.

Sarnia Tunnel depot looking east. Canadian Customs were housed in the building at the left of the picture. The Canadian depot was called Sarnia Tunnel to distinguish it from the former Great Western Sarnia depot located on the waterfront downtown.

– Pesha photo, George Smith Collection

Port Huron depot looking west, circa 1910.   – Pesha photo, George Smith Collection

The Grand Trunk operated the St. Clair Tunnel Company as an independent subsidiary until 1923. That year the bankrupt Grand Trunk was absorbed into Canadian National Railways and all Grand Trunk properties became part of CNR. The lines in Ontario and Quebec became Canadian National, while those in Michigan, Indiana and Illinois became Grand Trunk Western. They had operated for some time as Grand Trunk-Western Lines, but now the name became official. The St. Clair Tunnel Company continued as an independent subsidiary of its new parent company. To outsiders, the only change noticeable was the renumbering of locomotives from 2655-2660 to 9150-9155. Operations remained essentially the same, with steam road power being cut off at Sarnia or Port Huron and the electrics hauling trains through the tunnel.

By the 1920s traffic and train sizes had increased to the point where the St. Clair Tunnel Company was running four locomotives per train. This created a shortage of motive power. To remedy the situation the company ordered one new unit from Baldwin-Westinghouse in 1926. This locomotive, #9156, delivered in January 1927, was a virtual copy of the originals purchased in 1907. They also acquired, second-hand, two locomotives from the Chicago, South Shore & South Bend Railroad. These were numbered 9175 and 9176. With nine locomotives on the roster, they were able to operate two sets of four, with the odd engine out for servicing on a rotating basis.

In 1941 Canadian National purchased a gasoline-powered locomotive from the National Harbours Board in Montreal. It was numbered 15707 and assigned to the St. Clair Tunnel Company for use as a line car. Prior to this, repair crews used one of the electric locomotives to position the work train when servicing the overhead wire. The use of a self-propelled locomotive improved routine service work in a couple of ways. Crews no longer had to communicate with the dispatcher to have the power turned off and on when they desired to reposition the work train. The power could be shut off at the beginning of a repair job and the work cars repositioned at will until the job was completed, saving much time. The dispatcher then ordered power restored upon notification that the repairs were complete. Also, crews no longer had to worry about damaging the wires by careless resetting of the pantograph on the top of the locomotive after a repair had been completed. (These were raised and lowered by compressed air and could easily damage the wire while being raised if care was not taken.)

Engine 15707 was later renumbered 707. It had its gasoline engine replaced with a diesel in 1950. It outlived the electrics, being used to remove the overhead wire on the St. Clair Tunnel before being returned to CNR ownership. It was transferred to perform the same duty for other electric lines which were being abandoned by CNR at the time. By the time the 1960s rolled around, it too had gone, the last vestige of the St. Clair Tunnel equipment.

The year 1958 was a transitional one for the St. Clair Tunnel Company. Two events took place that had a major impact on the tunnel. The first was a financial and corporate restructuring. The St. Clair Tunnel Company ceased to exist as a separate entity. Under order in council 1958-302, the St. Clair Tunnel Company entered into an agreement on February 18, 1958, with its parent company, Canadian National Railways, and on March 31, 1958, it

Gas electric #15707 at Sarnia, July 25, 1947.

– Paterson-George Collection

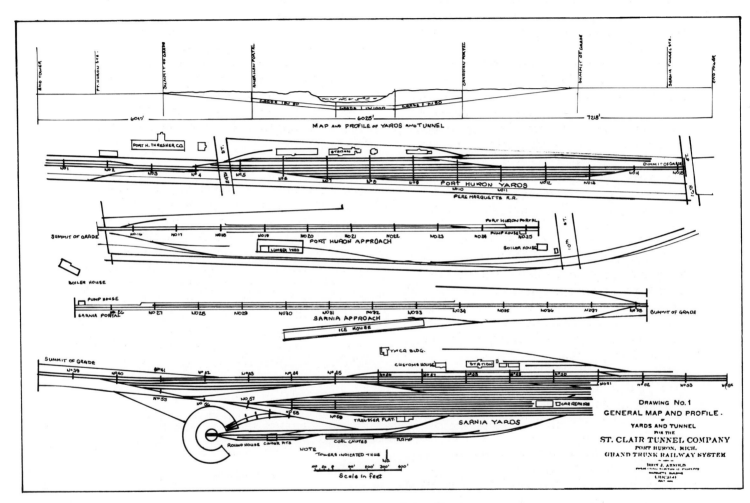

Map of Sarnia and Port Huron yards showing extent of electrification.   – St. Clair County Library

was formally amalgamated into CNR. The St. Clair Tunnel Company was now the St. Clair Tunnel Subdivision of the Great Lakes Region, Canadian National Railways.

The second event, which was by far the more noticeable, was the abandonment of the electric locomotives. The electrics, like the steamers before them, were destined to be replaced by diesels. They were victims not so much of improved technology as of old age and high maintenance costs. In later years CNR had purchased the foundry patterns and jigs from Westinghouse and were manufacturing their own repair parts for the engines. This costly process, combined with the dieselization of both CNR and Grand Trunk Western, sounded the death knell for the ancient engines. With improved ventilation, the diesels were able to run through the tunnel, saving the time required to change engines and crews at both ends. Only one change of engines was done, usually at Port Huron, where GTW would take over westbound trains from the CNR and deliver eastbound trains to them.

On September 28, 1958, the electrics hauled their last train through the St. Clair Tunnel. By the following spring the overhead wire would be gone and the locomotives themselves cut up for scrap.

Electrification of the St. Clair Tunnel had been extremely successful. From the crew's and passengers' standpoint, the noxious fumes and smoke had been eliminated, making a trip through the tunnel much more pleasant. From the company's standpoint, the electrification was a financial success. The increased capacity eliminated the need for a second tunnel. Indeed, the double track which ran to within a few feet of the portal on either end of the tunnel was eventually removed. The capability to move longer trains reduced the number of trains while increasing the tonnage through the tunnel. The fact that the electric locomotives could stand in the tunnel for an indefinite time, should the need arise, changed routine operating methods. For the first time, air brakes could be used in the tunnel. This gave the engineers better control of their trains, resulting in fewer broken couplers and generally lighter equipment wear.

With electrification came the end of the caboose on the St. Clair Tunnel run. Rear-end crews now had room to ride in the spacious cabs of the locomotives. Since there was no caboose at the end of the train, the dispatchers at the summit towers had to devise a different method of determining if a train had parted in the tunnel. A lantern with a red globe was hung on the ladder of the last car of the train as it entered the tunnel. When the operator at the other end saw the lantern, he knew the train was complete. To verify this, the reporting marks of the last car were telegraphed to the opposite operator. These were checked as the train emerged from the tunnel.

Grand Trunk Western engine #4932 emerges westbound from the Port Huron end of the tunnel in May 1961.

— Don McQueen

CHAPTER FIVE:

Current Status

The use of diesel locomotives in the St. Clair Tunnel was inevitable. North American railroads began converting to diesel power in the 1930s. By the 1950s most major U.S. roads were completely dieselized. Canadian railroads were five to ten years behind their U.S. counterparts, but by the end of the 1950s they, too, had abandoned steam. The Grand Trunk Western, following the lead of its parent company, Canadian National Railways, was one of the last U.S. lines to dieselize.

The conversion from electric to diesel power in the tunnel was the result of two major factors. First, the electric plant and locomotives were approaching 50 years old. Maintenance and upkeep were becoming increasingly expensive. The equipment was obsolete and many replacement parts had to be custom fabricated, consuming much time and capital. The second contributing factor was the versatility of the diesel locomotive itself. Unlike an electric locomotive, which must take its power from an outside source by means of, in the case of the St. Clair Tunnel, an overhead wire, a diesel locomotive is free to travel anywhere on the railway. Thus engines could be used to move trains out on the line when not required to work the tunnel, eliminating the need for specialized locomotives.

This flexibility allows the railway to utilize any motive power available at the time they wish to send a train through the tunnel. Also, the capability to operate diesels of varying horsepower together means the power can be matched precisely to the size of the train.

To remove the diesels' exhaust Canadian National renovated the ventilation system, which hadn't been needed since the steamers were retired in 1908. New ducts were installed along each side of the tunnel and new blowers at the portals.

In 1949 CN had lowered the track in the tunnel to allow larger modern freight cars to pass. Even so, by the late 1960s cars such as tri-level auto racks and 85-foot auto-parts boxcars were unable to use the tunnel. These cars were being crossed at Detroit/Windsor by way of the CN's ferry service. This was an inconvenient situation for the railway servicing the auto plants in the Flint, Michigan, vicinity. Consequently a new car-ferry service was inaugurated between Port Huron and Sarnia. With flags flying and whistles blowing, the tug *Phyllis Yorke* went into operation on March 11, 1971, pushing the barge *St. Clair*. A few years later the tug *Margaret Yorke* and barge *Scotia II* were brought from Windsor to increase capacity at the Sarnia–Port Huron crossing. Both barges were cut down from powered ferries. The *St. Clair* is the former *Pere Marquette #12*, built in 1927 at Manitowoc, Wisconsin. She spent most of her life running between Windsor and Detroit. The *Scotia II* was built in England and operated on the Strait of Canso in Nova Scotia.

Having replaced the original car-ferry operation with the tunnel 80 years before, it is ironic that the railway should come full circle with this new

East-bound laser train exits the Canadian portal, 1987.
– from *Survivals* by Dianne Newell and Ralph Greenhill, published by The Boston Mills Press, 1989

The tug Margaret Yorke *and barge* Scotia II *with oversize multi-level automobile carrier and Union Pacific high-cubic-capacity auto-parts car, Sarnia, 1987.*
– Ralph Greenhill

service. This time not as a replacement, but simply to augment the tunnel.

The year 1971 also saw the end of passenger-train service through the tunnel. For the first time in its 80-year history the St. Clair Tunnel would be a freight-only operation. Railways in Canada and the U.S. had been abandoning unprofitable passenger runs at every turn. The formation of the Amtrak national passenger service in the U.S. gave CNR/GTW the opportunity they needed to drop the Toronto–Chicago train. This freight-only status would last until October 31, 1982, when Amtrak and VIA Rail, the Canadian passenger-train service formed in 1976, would join together to reinstate the Toronto–Chicago service.

In 1985 Canadian National began operation of its "Laser" train. This train consists of specially built, articulated cars with recessed wells to hold the containers. Unlike conventional trains which change engines at Sarnia and Port Huron, the "Laser" train operates between Montreal and Chicago with a crew change only as required.

As the St. Clair Tunnel heads into its second century, it is still an important link in international rail traffic between Canada and the United States.

Sarnia Portal showing duct work for ventilation system installed to handle diesel exhaust fumes.
– Lambton County Library

Camel back 0-10-0T #599 as delivered in 1891. The four steamers, numbered 598-601, were built by Baldwin Locomotive Works of Philadelphia. — Lambton County Library

CHAPTER SIX:

Equipment

Due to its very nature, the St. Clair Tunnel has had an unusual and varied stable of equipment during its 100 years of operation. During the construction of the tunnel, motive power, mainly 4-4-0's and small 2-6-0's, was supplied by the Grand Trunk Railway.

The first locomotives to carry the St. Clair Tunnel Company name were four 0-10-0T steamers numbered 598-601, built in February 1891 by the Baldwin Locomotive works of Philadelphia, Pennsylvania. They were built as camelback engines to facilitate bi-directional operation, eliminating the need to be turned at each end of their run. As delivered, they carried 195,000 pounds on their 50-inch drivers, making them the heaviest engines constructed at the time. This weight was reduced to 173,000 pounds when the water tanks and coal bunkers were removed about 1898. Cylinders were 22 by 28 inches. The firebox was 11 feet long and 3 feet 6 inches wide. The 74-inch-diameter boilers held 281 2¼-inch-diameter iron tubes. Steam pressure was 160 pounds per square inch. With a tractive effort of 58,500 pounds at the drawbar, they were capable of hauling a 760-ton train up the 2-percent grade of the tunnel. The centre drivers had blind tires to allow these large engines to negotiate the curves and switches normally accustomed to seeing small 4-4-0's and moguls.

The Tunnel Steamers

Weight on drivers (as built)	195,000 pounds
Rigid wheelbase	18 feet 5 inches
Tank capacity	1,800 gallons water
	3 tons coal
Cylinders	22 x 28 inches
Drivers	50 inches
Driver centres	44 inches
Tires (standard Otis steel)	3 inches
Tires (1st, 2nd, 4th & 5th pairs)	5½ inches (flanged)
(3rd pair)	6 inches (blind)
Boiler	74-inch diameter
	⅝-inch thick
Normal operating pressure	160 psi
Firebox	132 ½ by
	42⅛ inches
Tubes (281) iron	2¼ inches by
	13 feet 6 inches
Tractive effort	58,500 pounds

The appearance of the 0-10-0's was changed in the late 1890s, when tenders were added to give them more running time between fuel and water stops. In 1898 the engines were renumbered 1301-1304 and designated as class G. Their side tanks were removed at this time. In 1908 they were downgraded to yard switchers when they were replaced by the electrics. Their assignments are not certain, but it is thought that one engine worked Sarnia Yard, another Port Huron, and possibly two went to Fort Erie, Ontario. Apparently the two engines sent to Fort Erie didn't stay long. Their heavy weight, combined with their long, rigid wheelbase, had a detrimental effect on the switches in the yard, and they were either reassigned or scrapped.

At least two engines, #1301 and #1303, had their cabs moved to the rear of their boilers to accommodate switching. This change drastically affected their appearance, making them look more like conventional 0-10-0's. Number 1303 was relettered Grand Trunk. They were renumbered 2650-2653 in 1910, although they may never have actually carried those numbers. Scrapping began in 1916, with #2652 being the first to feel the scrapper's torch at Hamilton, Ontario. By 1920 her three sisters had experienced the same fate.

The first six electric locomotives purchased by St. Clair Tunnel were built in 1907 by Baldwin Locomotive Works-Westinghouse. These went into service May 17, 1908, replacing the steam locomotives. Designated class G1, they were double-ended box-cab types, and as delivered, carried numbers 1305-1310. The Grand Trunk renumbered them 2655-2660 in 1910. Upon Canadian National takeover in 1923, they were renumbered 9150-9155 and reclassified Z2a. In November 1949 they were renumbered a final time, 150-155, to make room for incoming diesels. While numbered as individual units, the St. Clair Tunnel originally considered them as three locomotives each consisting of two half units. Each half unit rode on six 62-inch spoked drive wheels and was powered by three 250-horsepower single-phase motors. The 3,300 volts AC line power was reduced to 240 volts or lower, by taps taken from a transformer, to control the speed. The controllers had 21 points giving a gradual increase in speed and allowing the engineers to maintain a constant drawbar pull. The transformer and motors were cooled by an electrically driven blower, drawing cooling air in through the shutters on the side of the cab.

Compressed air was provided by a 2-cylinder motor-driven air pump with a capacity of 45 cubic feet per minute. This supplied air to the brakes and main controller as well as the bell, whistle, sanders, and for raising and lowering the pantograph.

Two of the units were capable of hauling a 1,000-ton train through the tunnel at 30 miles per hour. Each unit had a nominal drawbar pull of 50,000 pounds, but during tests at Westinghouse's plant in Pittsburgh, Pennsylvania, the dynamometer showed in excess of 43,000 pounds on one half unit, or 86,000 pounds for a "complete" locomotive.

After the Canadian National takeover in 1923, traffic increased through the St. Clair Tunnel. In 1926 a repeat order was placed with Baldwin-Westinghouse for one locomotive. Number 9156 was delivered in January 1927. Classified Z2b, it varied from the six purchased 20 years earlier only in minor details. It did, however, weigh 6 tons more. Along with the other SCT engines, it was renumbered (#156) in November 1949 to make room for the new CN diesel numbering system.

In March 1927 the SCT acquired two second-hand units from the Chicago, South Shore & South Bend. These locos, the heaviest on the SCT at 71½ tons each, had been built in July 1916 for the Chicago, Lake Shore & South Bend, the predecessor of the CSS&SB. Numbered 9175-9176 and classified Z3a, they, too, were box-cabs. Originally designed to operate on 6,600 volts, they were converted to 3,300 volts. They were rated at 700

*Two of the original electrics at Sarnia, November 3, 1914. They carried numbers in the
2650's from 1910 until 1923.*

— NAC

Number 15707 and line cars #12 and #10 at Port Huron, July 1944. The unmarked car at the right may also have belonged to the Tunnel Company.   – Paterson-George Collection

horsepower and produced a continuous tractive effort rating of 13,585 pounds.

The interior arrangement was reversed from that of the original Baldwin-Westinghouse boxcabs, having the electrical equipment located in the centre of the body and a walkway around the perimeter. They also rode on B-B trucks with 50-inch-diameter wheels. Renumbered 175-176 in November 1949, they were withdrawn from service in 1958 and scrapped along with the other SCT electrics in April 1959.

The most unique piece of motive power on the SCT was gas-electric #15707. This locomotive had been built by English Electric in 1928 as #44, a tower car for the National Harbours Board. It was purchased in 1941 by Canadian National Railways and renumbered SCT #15707. It was transferred to SCT operations and renumbered SCT #707 in 1950. In the early 1960s it was renumbered CNR #15707 for the second time. It spent its life on the SCT as a line car for servicing the overhead wire. Originally equipped with a Leyland 6 cylinder gasoline engine, it was repowered in 1950 with a GM model 6-71 diesel engine. When SCT electric operations ended, #707 was used to take down the wire. It then performed similar duties on other CN electric lines being abandoned in the late 1950s and early '60s. It was finally retired on June 5, 1968, and scrapped at Montreal.

Since the abandonment of electric operation, service through the tunnel has been provided by CN and GTW diesels, and more recently VIA Rail and Amtrak on passenger trains.

Usually the least chronicled equipment on any railway is the rolling stock. And so it is on the St. Clair Tunnel. Due to the tunnel's role as a transfer operation between two divisions of the same railway, the SCT owned no interchange cars. All its rolling stock was for either construction or service use.

Poor's Manual of Railroads, in its first listing of the SCT in 1893, records 4 steam locomotives and 25 dump cars. These are probably the four-wheel dump cars which appear in some photographs taken during the construction of the tunnel. These cars carried Grand Trunk reporting marks 12401-12650. They were 8 feet long with 40,000 pounds capacity. Little else is known about when these cars were built or who built them. They no longer appeared in *Poor's Manual* as of 1901. Whether they were scrapped or returned to Grand Trunk ownership is not known. It should be noted that the listings in *Poor's Manual* consistently appear two years behind, so these cars could have been removed from SCT registry as early as 1899.

It took only a few years for the SCT to discover that conductor's vans were needed. Nine vans were obtained in the late 1890s. They are first listed in *Poor's Manual* for 1898. Numbered 1-9, they appear to have dimensions similar to Grand Trunk cabooses, although they share few details in common. The body has three windows, as opposed to the usual two on a Grand Trunk van of the period. The cupola is unlike anything else on the Grand Trunk, appearing to have no side windows. Even the steps don't follow Grand Trunk standard design. It is possible they were constructed by the Grand Trunk at the car shops in Port Huron. Disposition of these vans is uncertain, although one is believed to have ended its life as a yard office at Sarnia.

The number of vans remaining dropped to one in the 1919 *Poor's Manual*, and they were completely gone from the 1922 listing. The use of electric locomotives with cabs large enough to house the entire train crew had made the vans unnecessary. There is a commonly held misconception that during the steam era the vans were operated directly behind the engines for safety reasons. The reason suggested is that if the train were to part in the tunnel, the rear-end crew would not be left in the noxious fumes. This was obviously not the situation in the three disastrous asphyxiation incidents. Also, no photos have come to light to show a freight train with a caboose coupled between the locomotive and the train.

The advent of electric power on the SCT brought with it the requirement for cars to work on the catenary. Line cars #10 and #12 first appear listed in the January 1915 GTR equipment list. They were converted from old boxcars, with platforms added to their roofs for work crews to stand on. Number 12 appears to be a standard GTR wooden, truss-rod boxcar of the type built after the turn of the century. These cars were 36 feet long, 8 feet 6 inches wide and 8 feet high inside. Number 10 was an older wooden car, possibly 33 to 34 feet long and 7 feet high, built sometime before 1900. The 1920 *Poor's Manual* lists a third work car. Whether this was another converted boxcar, and what the number was, is not known.

The most unusual piece of rolling stock used in the St. Clair Tunnel was a former Canadian National boxcar constructed in October 1927. After the overhead wires were removed in 1959, CN officials wanted to know the maximum size car that could pass through the tunnel. Workers at Sarnia car shops attached steel plates to each end of the car at the roof line. These were contoured to match the in-

side of the tunnel. To these were fastened pieces of pipe which held chalk markers. As the car was run through the tunnel, the chalk was raised until it touched the inside of the lining, indicating the limits restricting the use of oversize cars. After this task was accomplished, the car spent the remainder of its life being used to distribute sand around the switches in Sarnia Yard during winter. It carried number 70049 in company service and was scrapped in the late 1980s.

To ferry men and materials across the river, during construction the St. Clair Tunnel Company purchased a small steamboat, the *W.J. Taylor*. It may have been purchased early enough to have been used in the taking of core samples during testing of the river bed. The *Taylor* was built in Chatham, Ontario, in 1883. It was 35 feet long by 8 feet beam, with a depth of 3 feet. Power came from a 10-horsepower steam engine which drove a screw. It was probably purchased second-hand, since the tunnel company was not formed until 1884 and did not begin to take core samples until 1885. The SCT retained the steamboat's original name and its home port of Chatham. Although unconfirmed, it is thought that the boat may have been named after the son of the founder of T.H. Taylor Co. Ltd. (currently Taylor Grain Ltd.) of Chatham. The Taylors were a prominent Chatham family with numerous financial interests in the city. W.J. Taylor was born on November 27, 1863. He eventually became vice-president of T.H. Taylor Co. Ltd.

Final disposition of the *W.J. Taylor* is not known. It was still registered to the SCT as late as 1913 and was probably removed from registry sometime within the next three or four years.

The W.J. Taylor *was used to ferry men and materials across the river during construction of the tunnel. She was built in Chatham, Ontario, in 1883.*

– Burton Historical Collection, Detroit Public Library

St. Clair Tunnel Locomotive Roster												
Original #	RE# 1898	RE# 1910	RE# 1923	RE# 11/49 1949	Builder	Const. #	Blt. Date	GTR Class	CNR Class	Disposi-tion	Wheel Arrange-ment	Notes
598	1301	2650	–	–	BLW	11586	2/91	G	–	Scrapped 5/20	0-10-0T	a
599	1302	2651	–	–	BLW	11589	2/91	G	–	Scrapped 5/20	0-10-0T	a
600	1303	2652	–	–	BLW	11590	2/91	G	–	Scrapped 1/16	0-10-0T	a
601	1304	2653	–	–	BLW	11595	2/91	G	–	Scrapped 8/20	0-10-0T	a
–	orig. # 1305	2655	9155	155	BLW/ WEST	29993	1/07	G1	Z2a	Scrapped 4/59	c	
–	1306	2656	9154	154	BLW/ WEST	31823	9/07	G1	Z2a	4/59	c	
–	1307	2657	9153	153	BLW/ WEST	31870	10/07	G1	Z2a	4/59	c	
–	1308	2658	9152	152	BLW/ WEST	31924	10/07	G1	Z2a	4/59	c	
–	1309	2659	9151	151	BLW/ WEST	31871	10/07	G1	Z2a	4/59	c	
–	1310	2660	9150	150	BLW/ WEST	32851	7/08	G1	Z2a	4/59	c	
–	–	–	orig. # 9156	156	BLW/ WEST	59767	1/27	–	Z2b	4/59	c	b
–	–	–	9175	175	BLW/ WEST	43681	7/16	–	Z3a	4/59	b&b	c
–	–	–	9176	176	BLW/ WEST	43682	7/16	–	Z3a	4/59	b&b	c
–	–	–	15707 –	707	English Electric	734	/28	–	–	6/5/68	b&b	d

Notes:
a) Built as sidetank/camel backs; tenders added before 1897; side tanks removed c.1898, cabs moved to rear c.1910.
b) purchased new January 1927.
c) purchased used March 1927; built as Chicago, Lakeshore & South Bend (later Chicago, South Shore & South Bend) #505 and #506
d) built for National Harbours Board as #44; purchased by CN 1941, assigned to SCT; returned to CNR c.1959 re #CN 15707

Left side of #1304, circa 1900.

— NAC

Right side of #1304, probably during the winter of 1907-08. — NAC

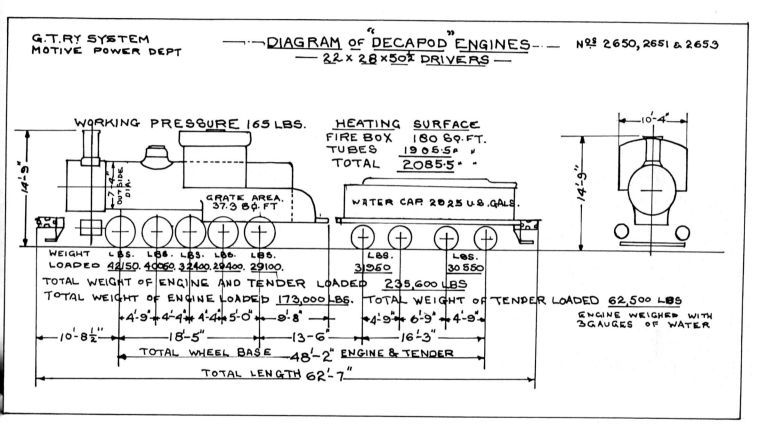

Diagram of 0-10-0's in their final configuration, taken from GTR motive-power book. Note the boiler diameter is mistakenly shown as 7' 4" instead of 74".

— CNR

SUB CLASS	DATE BUILT	BUILDER	BUILDER'S ORDER No	BUILDER'S No	PREVIOUS ROAD No AND INITIALS	PRESENT ROAD No
Z-2-a	1907	B.& WEST.H		32851, 31874, 31924, 31870, 31823, 29993	2655 to 2660 ST C.T.Co 9160 to 9165	150 to 155
Z-2-b	1927	" "	32592	59767	9156 ST C.T.Co	156

ST. CLAIR TUNNEL CO.
TYPE ELECTRIC CLASS Z-2

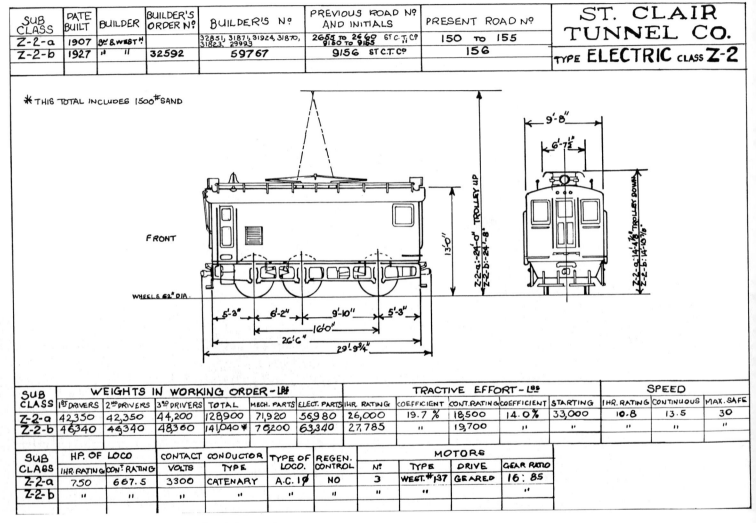

* THIS TOTAL INCLUDES 1500# SAND

FRONT

WHEELS 62" DIA

13'-0"

9'-8"
6'-7½"

Z-2-a: 24'-0" TROLLEY UP
Z-2-b: 24'-8"

Z-2-a: 14'-4⅜" TROLLEY DOWN
Z-2-b: 14'-10⅞"

5'-3"   6'-2"   9'-10"   5'-3"
26'-6"
16'-0"
29'-9¾"

SUB CLASS	WEIGHTS IN WORKING ORDER – LBS							TRACTIVE EFFORT – LBS				SPEED		
	1ST DRIVERS	2ND DRIVERS	3RD DRIVERS	TOTAL	MECH. PARTS	ELECT. PARTS	1HR. RATING	COEFFICIENT	CONT. RATING	COEFFICIENT	STARTING	1HR. RATING	CONTINUOUS	MAX. SAFE
Z-2-a	42,350	42,350	44,200	128,900	71,920	56,980	26,000	19.7 %	18,500	14.0%	33,000	10.8	13.5	30
Z-2-b	46,340	46,340	48,360	141,040 *	76,200	63,340	27,785	"	19,700	"	"	"	"	"

SUB CLASS	HP. OF LOCO		CONTACT CONDUCTOR		TYPE OF LOCO.	REGEN. CONTROL	MOTORS			
	1HR. RATING	CONT. RATING	VOLTS	TYPE			No	TYPE	DRIVE	GEAR RATIO
Z-2-a	750	667.5	3300	CATENARY	A.C. 1Ø	NO	3	WEST. #137	GEARED	16: 85
Z-2-b	"	"	"	"	"	"	"	"	"	"

Diagram of Z2-class electric locomotives.

— CNR

Electric #9156 as delivered from Baldwin-Westinghouse in 1927. — CNR

Electric #9175 shortly after its purchase from the Chicago, South Shore & South Bend in 1927.

– Bob Gray photo

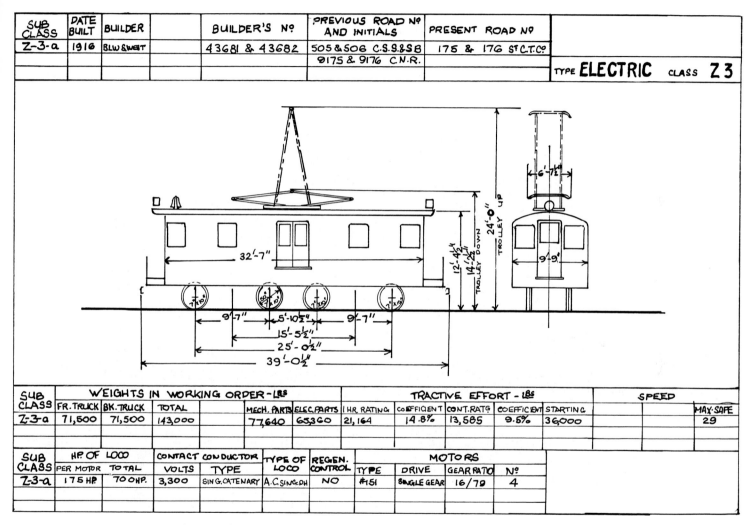

SUB CLASS	DATE BUILT	BUILDER		BUILDER'S Nº	PREVIOUS ROAD Nº AND INITIALS	PRESENT ROAD Nº	
Z-3-a	1916	BLW & WEST		43681 & 43682	505 & 506 C.S.9. & S B	175 & 176 St C.T.Cº	
					9175 & 9176 C.N.R.		TYPE **ELECTRIC** CLASS **Z 3**

SUB CLASS	WEIGHTS IN WORKING ORDER - LBS						TRACTIVE EFFORT - LBS					SPEED	
	FR.TRUCK	BK.TRUCK	TOTAL		MECH.PARTS	ELEC.PARTS	1 HR RATING	COEFFICIENT	CONT.RATG	COEFFICIENT	STARTING		MAX·SAFE
Z-3-a	71,500	71,500	143,000		77,640	65,360	21,164	14.8%	13,585	9.5%	36000		29

SUB CLASS	HP. OF LOCO		CONTACT CONDUCTOR		TYPE OF LOCO	REGEN. CONTROL	MOTORS					
	PER MOTOR	TOTAL	VOLTS	TYPE			TYPE	DRIVE	GEAR RATIO	Nº		
Z-3-a	175 HP	700 HP.	3,300	SING. CATENARY	A.C. SING. PH	NO	#151	SINGLE GEAR	16/79	4		

Diagram of Z3-class electric locomotives.

— CNR

Electric #175, sometime in the 1950s, displays its final paint scheme. – Paterson-George Collection

Electrics #151 and #150 about to enter the Canadian end of the tunnel. All the electrics were renumbered in 1949 to make room for diesels in CNR*'s numbering system.* — CNR

Builder's photo of gas-electric #15707 (later #707) in 1928.   – Ray Corley Collection

Number 15707 was renumbered #707 in 1949. Here she wears her final St. Clair Tunnel Company livery sometime in the 1950s.
— Paterson-George Collection

St. Clair Tunnel Company caboose #1 parked in front of Sarnia Tunnel depot, circa 1908. - NAC

Chronology

1858	Great Western Railway arrives at Sarnia.	July 1887	Test shaft abandoned by Sooysmith.
1859	Grand Trunk Railway arrives at Point Edward.	April 20, 1888	GTR begins to dig full-size tunnel. Work abandoned almost immediately.
November 1859	GTR leases Chicago, Detroit & Canada Grand Trunk Junction Railway. Connection between Point Edward and Fort Gratiot is made by means of a "swing ferry."	May 7, 1888	GTR begins second set of test borings.
		January 1889	GTR begins excavation to put shields in place.
1872	Steam-powered car ferry *International* launched.	July 11, 1889	Shield tunnelling begins on U.S. side of river.
1875	Steam-powered car ferry *Huron* added to fleet.	September 21, 1889	Shield tunnelling begins on Canadian side.
1880	Chicago & Grand Trunk Railway organized.	August 23, 1890	Fifteen feet separate shields. An auger hole is bored and a plug of chewing tobacco is passed through—the "first freight" through the tunnel.
1882	Great Western absorbed into the Grand Trunk.		
1882	Walter Shanley hired by Grand Trunk to do survey for tunnel.	August 25, 1890	An opening is shovelled out and officers of the company pass through.
April 19, 1884	St. Clair Frontier Tunnel Company incorporated in Canada.	August 30, 1890	The two shields meet at 11:30 PM., out only ¼ inch.
1885	GTR makes set of test borings along route of tunnel.	April 9, 1891	Sarnia yard engine #253 makes first trip through tunnel.
October, 18, 1886	Port Huron Railroad Tunnel Company incorporated in Michigan.	September 19, 1891	St. Clair Tunnel formally opened.
		October 24, 1891	First freight train transits tunnel.
November 1886	Two companies merged to form the St. Clair Tunnel Company	December 7, 1891	First passenger train transits tunnel.
December 1886	Sooysmith and Company begin preliminary shaft.		

January 31, 1892	Conductor George Hawthorne becomes first victim of asphyxiation in tunnel.
November 28, 1897	Second fatal accident occurs in tunnel. Three men die of asphyxiation.
October 9, 1904	Six men die in third asphyxiation incident in tunnel.
1905	Grand Trunk signs contract with Westinghouse Company to electrify St. Clair Tunnel.
February 20, 1908	First electric locomotive makes trip through tunnel.
May 17, 1908	Steam engines retired from tunnel.
November 12, 1908	Formal ceremonies celebrating new electric operation of tunnel.
1917	Sabotage attempt to blow up tunnel.
1927	St. Clair Tunnel Company receives one new and two used locomotives.
1941	Gas-electric locomotive #15707 added to roster.
1949	Engines renumbered to make room for new diesels being added to CNR roster in early 1950s.
February 1958	St. Clair Tunnel Company becomes St. Clair Tunnel Subdivision of Canadian National Railways.
September 28, 1958	Diesel locomotives replace electrics in tunnel operation. Electric locomotives are scrapped early the next year.
1971	CNR/GTW abandon passenger-train operation through St. Clair Tunnel.
March 1971	Tug and barge operation augment tunnel, moving oversize cars that cannot be taken through tunnel.
October 31, 1982	Amtrak/VIA reinstate passenger service between Toronto and Chicago by way of St. Clair Tunnel.
1985	CN begins operation of "Laser" train intermodal service between Chicago and Toronto/Montreal.

This car was used to determine the clearance which restricts the size of freight cars which can pass through the tunnel. It spent its last days as a sand car in Sarnia Yard.
– Gord Taylor photo

Facts About the St. Clair Tunnel

One of the longest submarine tunnels in the world.

Length of tunnel from portal to portal is 6,025 feet.

Length including approaches is 2.5 miles.

Length under the river bed is 2,290 feet.

Length of the American cut is 2,487 feet.

Length of the Canadian cut is 3,116 feet.

Interior diameter of tunnel is 19 feet 10 inches.

Grade on the approaches is 2-percent, or a rise of 1 in 50. The centre has a grade of .1-percent, or 1 in 1,000.

Weight of cast iron used in lining is 56 million pounds.

Cost of construction was $2,700,000.

One of the first tunnels in the world converted to electricity.

Length of electrified zone is 4 miles.

Cost of electrification was $500,000.

All movements were controlled from the tower at East Summit in Sarnia. An almost identical tower stood at West Summit in Port Huron. – Lambton County Library

Acknowledgments

To produce any historical book requires the assistance of many people. I would like to thank all those who have shared their time, artifacts and recollections with me over the past few years. Without their help this book could not have been written. I would like to thank the following for their contributions:

Fred Ashby; Dana Ashdown; Archives of Ontario; Gene Buell; Burton Historical Collection, Detroit Public Library; Canadian National Railways, J. Norman Lowe, Connie Romani; Ray Corley; Greg Degowski; Sandy Duffy; Tom Gaffney; Dick George; Bill Glassco; Mark Hanslip; Lambton County Library, Wyoming; Jack Lewis; Irene McLean; Don McQueen; National Archives of Canada; Al Paterson; Perth County Archives, Stratford; Glen Phillips; Port Huron Museum of Arts and History; Port Huron *Times-Herald*; Sarnia Historical Society; Sarnia *Observer*; Sarnia Public Library; Gary Shurgold; the late George Smith; Mike Smout; and Gord Taylor.

A special acknowledgement to Bob Gray, without whose encouragement this book would not have seen light. When I started researching this book, I was told, "Go see Bob Gray."

When I did, Bob opened his heart and his home to me. He gave me total access to his marvellous railroadiana collection and his wonderful memories from working on the St. Clair Tunnel and the Grand Trunk Western in Port Huron. Along about the second or third visit to Bob's home, we agreed that he should co-author this book, since the subject was so dear to him. Unfortunately fate has a way of intervening and our plans weren't realized. Bob passed away in March, 1988 at 87 years of age. We'll miss you, Bob.

I would especially like to thank the following for their input: Rich Chrysler and Gerry Elder, who proofread the manuscript and whose suggestions were invaluable; Lorraine Hodgins and Diane Wallace, who struggled through my handwritten notes to type the manuscript; and lastly John Denison and The Boston Mills Press for their support and encouragement.

Clare Gilbert
Parkhill, Ontario
September 1989

Index